SIMULATION OF REMOTELY SENSED RAINFALL FIELDS USING COPULAS

SIMULATION OF REMOTELY SENSED RAINFALL FIELDS USING COPULAS

Amir AghaKouchak

This monograph is based on a Doctoral dissertation submitted at the University of Stuttgart in partial fulfillment of the requirements for the Doctoral Degree in Civil & Environmental Engineering. The entire document appeared in Transactions of the Institute of Hydraulic Engineering, University of Stuttgart (ISSN 0343-1150), Volume 188.

Simulation of Remotely Sensed Rainfall Fields Using Copulas / Amir AghaKouchak.
Transactions of the Institute of Hydraulic Engineering
University of Stuttgart (ISSN 0343-1150)
Includes bibliographical references.
ISBN 978-3-933761-92-7 (pbk.)

To Nasrin

Acknowledgments

Completing this document has definitely been a challenge and could not have been accomplished without the assistance and support of many individuals and institutions. My deepest gratitude goes to Prof. András Bárdossy for his detailed criticism and unyielding support. His profound knowledge and expertise in this area of research has always led me throughout my work. I am also thankful for the knowledge I gained while attending his doctoral classes and conducting research with him. I would like to thank Prof. Uwe Haberlandt, Prof. Rainer Helmig and Prof. Markus Friedrich for their excellent feedback that helped me broaden my research perspectives.

I am pleased to acknowledge that a major portion of the funding for this study was provided by a fellowship from the Department of Civil Engineering, University of Louisiana at Lafayette, USA, through a Research Development grant for uncertainty analysis of radar rainfall estimates. Without this financial assistance this study could not have been undertaken. I would like to express my gratitude to Dr. Habib, Department of Civil Engineering, University of Louisiana at Lafayette, who challenged my ideas and my writing with his sharp comments. His knowledge of weather radar systems has been invaluable, and I greatly appreciate his assistance and guidance. Without his encouragement, effort and advice, this work would not have been realized. I also want to praise Dr. Jing Li for her extraordinary advice and willingness to share her knowledge and experience. I am deeply indebted to Prof Soroosh Sorooshian, Director of the Center for Hydrometeorology and Remote Sensing, University of California Irvine for his support at the last stages of my work. I would like to take this opportunity to thank the National Oceanic and Atmospheric Administration (NOAA) Educational Partnership Program award that allowed me to meet many students and researchers. The program played a significant role in sparking my interest in remote sensing research.

I am very grateful to Jeff Tuhtan, William Roth and Amy Henschke who have given so generously of their time reviewing this document and offering editorial suggestions for improvement. I offer special thanks to Manuel Lorenz for all the time he spent on writing the German version of the summary of my dissertation. I would like to thank my friends and colleagues who have provided support in a variety of ways. These individuals include Mahyar Mehdizadeh, Jeff Tuhtan, Krista Uhrmann, Min Zhang, Nasrin Nasrollahi, Thomas Pfaff, Dr. Jochen Seidel, Claudia Hojak, Dr. Matthias Schneider, Dr. Marx, Aaron LaFramboise, Dr. Riegger, Diane Hohnbaum, Dr. McManis, Prof. Kobus, Dr. Ciach, Dr. Thapa, Amin Darbandi, Marieh Zargar, Ehsan Rabiei, Henning Lebrenz and Dr. Scharf. Furthermore, thanks to all the faculty members and colleagues at the Institute for Hydraulic Engineering, University of Stuttgart and the Department of Civil Engineering, University of Louisiana at Lafayette for the overall good working atmosphere. Your contribution is appreciated.

Finally, I am extremely grateful to my beloved family, my parents and my wife Nasrin, for their unconditional love and support over the last few years. To Nasrin, whose extraordinary faith in my abilities, ceaseless encouragement and emotional support inspired me to complete my studies and this research. She has made many sacrifices to assist me with my work on this project. Her love and support of my endeavors can never be repaid. Thank you for so graciously volunteering all the extra effort and for never complaining about the weekends I had to work.

AMIR AGHAKOUCHAK
November 2009

"My words are unable to express the mystery of love,
beyond the limit of narration, is the explanation of longing."
Hafez 1315-1390 A.D.

CONTENTS

List of Figures

List of Tables

List of Symbols and Acronyms

Symbols

A	multiplicative factor in the $Z - R$ relationship
b	exponent in the $Z - R$ relationship
C	copula
C^n	n-dimensional copula
c	copula density functions
F^n	n-dimensional multivariate function
F_n^{-1}	inverse multivariate distribution function
k	v-copula parameter
m	v-copula parameter
n	dimension of the copula
$N^n_{\mu,\sigma,\rho}$	multivariate Gaussian copula
n_{out}	number of time steps that rain gauge measurements do not fall within of the simulated ensemble

O_i	observed values
P_i	precipitation (multi-sensor precipitation estimates)
r	range
R_{gauge}	surface rainfall data
R_{radar}	radar estimates
R_i	radar rainfall
Ran	Range
rho_s	Spearman rank correlation
s	sill
s_0	nugget effect
S_i	simulated values
t^n	multivariate Student CDF
t_ν	univariate Student distribution with ν degrees of freedom
Z_i	Reflectivity
χ_i	rank of x_i in X
ϵ	error
ϵ_1	proportion error
ϵ_2	purely random error
λ_{lo}	lower tail
λ_{up}	upper tail
ν	degrees of freedom
ρ	correlation matrix
σ	standard deviation
φ	probability density function of the normal distribution
$\xi(d)$	positive-definite correlation function
ξ_i	rank of y_i in Y

Acronyms

ARS	Agricultural Research Service
ADE	Alternating Direction Explicit
ARMA	Auto-Regressive Moving Average

ECDF	Empirical Cumulative Distribution Function
DEM	Digital Elevation Model
EOS	Earth Observing System
FORTRAN	Formula Translator Programming language
GLM	Generalized Linear Models
GC	Goodwin Creek watershed
GPM	Global Precipitation Mission
GSSHA	Gridded Surface Subsurface Hydrologic Analysis
IFM	Inference Function for Margins
LOOCV	Leave-One-Out Cross-Validation
LW	Little Washita watershed
MLE	maximum likelihood estimator
MAE	Mean Absolute Error
MPE	multi-sensor precipitation estimates
CDF	cumulative distribution function
NSF	National Science Foundation
NWS	National Weather Service
NEXRAD	Next Generation Weather Radar
E-Model	Random Error Model
RFC	River Forecast Centers
RMSE	Root Mean Square Error
TRMM	Next Generation Weather Radar
USACE	US Army Corps of Engineers
ERDC	Engineering Research and Development Center
USGS	US Geological Survey
VPR	Vertical Profiles of Reflectivity
WSR-88D	Weather Surveillance Radars - 1988 Doppler

Abstract

Rainfall is a major input in hydrological and meteorological models. Quantification of rainfall and its spatial and temporal variability is extremely important for reliable hydrologic and meteorological modeling. Hydrological and climate studies have long relied on rain gauge measurements. While rain gauge measurements do not provide reasonable areal representation of rainfall, remotely sensed precipitation estimates offer much higher spatial resolution. Recent technological advances in the field of remote sensing have led to an increase in available rainfall data on a regional and global scale. However, the advantages of remotely sensed data are limited by complications related to the indirect nature of remotely sensed estimates. Previous studies confirm that remotely sensed rainfall estimates are subject to various errors, and for future use in hydrologic and climate studies, efforts are required to determine the accuracy of data and their associated uncertainties. Despite extensive research, however, uncertainties associated with remotely sensed rainfall estimates are not yet well quantified. Radar rainfall estimates, for example, are associated with several different error types that arise from various factors such as beam over-shooting, partial beam filling, non-uniformity in vertical profiles of reflectivity (VPR), inappropriate $Z - R$ relationship, spatial sampling pattern, hardware calibration and random sampling error. It is expected that uncertainties in rainfall input data will propagate into predictions from hydrologic and meteorologic models; therefore, accurate characterization and quantification of such errors in radar data and the induced uncertainties in hydrologic applications is an extremely important, yet challenging issue.

So far, a multitude of approaches and extensive research efforts have been undertaken to develop an uncertainty model for remotely sensed rainfall estimates. In order to assess rainfall uncertainties, one can simulate an ensemble of precipitation fields that consists of a large number of realizations, each of which represents a possible rainfall event that can occur. Subsequent runs of a hydrological or meteorological model using simulated ensembles of rainfall estimates would then allow an assessment of uncertainty propagation due to the precipitation input. One way to generate an ensemble of rainfall estimates is to stochastically simulate random error fields and impose them on radar estimates. This study intends to develop different stochastic techniques for simulation of radar-based rainfall fields through simulating random error fields and imposing them over remotely sensed rainfall estimates. Four different models are developed and discussed in this work. In the first and second models, two elliptical copulas, Gaussian and t-copula, are used to describe the dependence structure of radar rainfall error and to simulate multivariate rainfall error fields. In the third model, an asymmetrical v-transformed copula is employed for error simulations. In the fourth model, rainfall fields are generated by perturbing rainfall estimates with two normally distributed error terms: a purely random component and a component proportional to the magnitude of the rainfall rates. In the first three models, having described the dependencies using copulas, the empirical distribution function of observed rainfall error is numerically approximated and applied to the simulated error fields so that the simulated realizations are similar to those of the observed in terms of the distribution function. In the fourth model, however, the error is assumed to be normally distributed. In all the models, available observations of radar rainfall error (the differences between radar estimates and rain gauge measurements) are used to condition the simulated fields on observations.

In order to examine reliability and performance of the developed models, several case studies are presented over a small watershed in Mississippi, USA, and a large watershed in Oklahoma, USA. Both radar reflectivity data (Level II) as well as Stage IV Next Generation Weather Radar (NEXRAD) multi-sensor precipitation estimates are used as input to the models. The simulated rainfall fields obtained from different models are compared with original radar estimates with respect to statistical properties, extreme values and spatio-temporal dependencies. Moreover, a physically based model is used to demonstrate the application of the presented rainfall field generators in streamflow analysis. In subsequent chapters, after introducing the models, their strong and weak points are highlighted and discussed in detail. It is hoped that the results of this research can be used to assess the uncertainties associated with radar-based rainfall estimates; as it is believed, that with accurate information about surface rainfall and its associated uncertainties, hydrologists and meteorologists have the potential to improve hydrologic predictions and global climate studies.

CHAPTER 1

INTRODUCTION

1.1 MOTIVATION

Recent climate studies have highlighted the importance of water resources engineering and management with respect to future global decision making and risk analysis (IPCC (2007); Ruddiman (2005); Miller and Edwards (2001)). Predicting the available water resources for human consumption, agriculture and industry and quantifying the associated uncertainties are of particular interest. Understanding the underlying processes in the hydrologic cycle is fundamental to water resources management and climate studies. The water content in the atmosphere is in continuous interaction with the land surface through precipitation, evaporation from surface waters and transpiration from plants. Precipitation may also result in surface runoff, infiltration into the subsurface and groundwater recharge.

Hydrological and meteorological models, which are highly simplified representations of the natural hydrologic cycle, are often used to model the interconnected movement of water between the land surface, atmosphere and oceans. Reasonable management and decision making in the field of the water resources requires reliable measurements of weather variables that are crucial for modeling the hydrologic cycle. Precipitation is a major input in hydrological and meteorological models. It is governed by nonlinear physical processes and is also known to be highly variable in both space and time (Smith and DeVeaux (1992); Troutman (1983); Corradini and Singh (1985); Dawdy and Bergman (1969)). Reliable estimation of precipitation is essential for hydrologists, as it is believed that uncertainties associated with rainfall estimates will propagate in hydrologic modeling predictions. In fact, a very well founded hydrologic model without reliable and accurate estimates of precipitation may not be able to predict the output hydrograph. Traditionally, rain gauges have been the main source of surface precipitation measurements. However, rain gauges are sparsely spatially distributed and suffer from a lack of areal representation of precipitation, which can be quite limiting for hydrologic applications. Detailed information on surface precipitation at fine spatial and temporal scales is essential to drive land-surface hydrology models, to validate numerical weather prediction and climate models, and to perform a variety of hydrologic prediction and forecasting studies. In particular, hydrological and meteorological modeling over large scales requires high resolution precipitation data to capture the temporal and spatial variability of rainfall, which has been proven to affect the quality of hydrological predictions (Yilmaz et al. (2005); Goodrich et al. (1995); Shah et al. (1996); Faures et al. (1995); Hamlin (1983); Troutman (1983); Corradini and Singh (1985); Obled et al. (1994); Seliga et al. (1992); Dawdy and Bergman (1969); Rodda (1967)).

In recent years, the development of weather radar systems, satellites and remote sensing techniques have provided precipitation information at higher temporal and spatial resolutions than was previously possible from rain gauge measurements. Application of remotely sensed rainfall data in hydrological and meteorological predictions has increased significantly in the past few decades. Compared with rain gauge measurements, remotely sensed precipitation data provide higher spatial and temporal resolution. However, remote sensing data are subject to different types of error such as sampling uncertainty, inherent measurement and retrieval errors, among others. Radar estimates, in particular, are associated with several error types that arise from various factors such as beam over-shooting, partial beam filling, instrumental errors and uncertainties, non-uniformity in vertical profiles of reflectivity (VPR), inappropriate $Z - R$ relationship, spatial sampling pattern, hardware calibration and random sampling error (Hossain and Huffman (2008); Margulis et al. (2006); Dong et al. (2005); Hossain and Anagnostou (2005); Lucieer and Kraak (2004); Bastin et al. (2002); Crosetto et al. (2001)). Additionally, the weather may also affect radar rainfall observations (Steiner and Smith (2000)). For example, rainfall particles which are strongly distributed in one direction or thermodynamic state may influence radar estimates (Austin (1987); Zawadzki (1984); Battan (1973)).

Despite extensive research, the uncertainties associated with remotely sensed rainfall data have not yet been well quantified (Hossain and Huffman (2008); Margulis et al. (2006); Dong et al. (2005); Hong et al. (2006); Hossain and Anagnostou (2005); Lucieer and Kraak (2004); Bastin et al. (2002); Tian et al. (2009); Crosetto et al. (2001)). Austin (1987) mentioned that even for a set of high quality data, differences in the spatio-temporal resolution and area coverage of radar estimates and gauge measurements could be substantial. Similar conclusions have been reported by Kitchen and Blackall (1992); Anagnostou and Krajewski (1997); Lovejoy and Schertzer (1990); Ciach and Krajewski (1999)). Characterization and quantification of such uncertainties are extremely important, as it is believed that spatial and temporal variability in input rainfall is one of the main sources of error in rainfall-runoff processes and hydrologic predictions (Schuurmans and Bierkens (2007); Tetzlaff and Uhlenbrook (2005); Syed et al. (2003); Arnaud et al. (2002); Bell and Moore (2000); Winchell et al. (1998); Shah et al. (1996); Goodrich et al. (1995); Faures et al. (1995); Obled et al. (1994); Seliga et al. (1992); Hamlin (1983)).

One way to assess the spatio-temporal uncertainties of precipitation is to simulate an ensemble of precipitation fields that consists of a large number of realizations, each of which represents a possible rainfall event. Stochastically generated precipitation ensembles can then be used as input to hydrological and meteorological models to assess model prediction uncertainties. A great deal of effort has been put into developing stochastic models for simulation of multivariate rainfall fields (gauge, radar, satellite, etc.). Lall et al. (1996) introduced a nonparametric bootstrap approach to resample daily precipitation. Conditional resampling methods such as k-Nearest-neighbor bootstrap were employed in a number of studies to model multi-site rainfall data (e.g. Rajagopalan and Lall (1999) and Buishand and Brandsma (2001)). Using mixed exponential distributions for the simulation of nonzero amounts, Wilks (1998) introduced a Markovian stochastic model for simulation of daily precipitation. Following the Wilks approach, Khalilia and Leconte (2007) presented an algorithm to improve the efficiency of the model when the data are contaminated with noise. Haberlandt et al. (2008) proposed a two step stochastic simulation procedure for generation of space-time precipitation fields. In the first step an alternating renewal model is used for simulation of independent rainfall time series, whereby the model defines wet spell durations and intensities as well as dry spell duration. The dependencies between wet spell durations and intensities are described using 2-copulas. In the second step, a multivariate resembling technique is employed to account for the spatial dependence structure of rainfall fields. Hutchinson (1995) introduced a space-time model for rainfall data using the Auto-Regressive Moving Average (ARMA) method. Fowler et al. (2005) used a semi-Markov approach for simulation of multi-site monthly rainfall data. Kim et al. (2008) applied the Markov chain method to account for temporal dependencies and the direct acyclic graph method to describe spatial dependencies of daily rain gauge data. One limitation of multi-site rainfall simulation is that accounting for spatial and temporal dependencies simultaneously is not straightforward. Pegram and Clothier (2001) introduced the String of Bead model for radar rainfall simulation based on two auto-regressive time series, one at the image scale and the other at the pixel scale, assuming log-normal marginal distribution for radar rainfall at the pixel scale. Seed and Srikanthan (1999) modeled space-time radar rainfall data using a multi-fractal (multiplicative cascade) approach where each level of the multiplicative cascade was linked in time using a different ARMA(1,1) model. Clark et al. (2004a) and Clark et al. (2004b) proposed a non-parametric rainfall simulation approach whereby spatio-temporal variabilities were preserved by reshuffling the simulated ensemble to match the observed spatio-temporal characteristics. Chandler and Wheater (2002) presented a simple model based on Generalized Linear Models (GLM). Furthermore, a number of parametric models conditioned on other meteorological quantities have been introduced to simulate rainfall and other climatological variables (see Bardossy and Plate (1992), Bras and Rodriguez-Iturbe (1976) and references therein). One limitation of parametric models is that the number of parameters grows with an increase in the dimension of the model. For a complete review on different rainfall simulation models, interested readers are referred to Wilks and Wilby (1999) and Mehrotra et al. (2006).

An alternative approach to obtaining an ensemble of rainfall fields is to simulate error fields and impose them over observed rainfall data. Krajewski and Georgakakos (1985) introduced a two-dimensional non-stationary random field generator for daily radar error. The application of this model in hydrological modeling is limited due to its low temporal resolution. Germann et al. (2006) proposed that the radar estimates can be perturbed with stochastic fields in order to obtain an ensemble of radar estimates. In a recent study, Ciach et al. (2007) developed an operational approach based on empirical investigations of joint samples of radar and ground surface data whereby the radar rainfall uncertainties consist of a systematic distortion function and a stochastic component. The radar error components are then estimated using non-parametric estimation schemes. Based on the work presented by Ciach et al. (2007), Villarini et al. (2009) developed a radar rainfall field generator and a model to produce probability of exceedance maps of radar rainfall conditioned on given radar rainfall estimates.

It is well known that hydrological and meteorological data are dependent in both space and time. A reasonable and robust model for rainfall simulation is expected to produce rainfall fields with similar spatial and temporal dependencies to those of the observed. However, preserving spatial and temporal dependency of simulated rainfall fields provides a considerable challenge. So far, a number of models have been developed to generate spatially correlated random fields using covariance or a variogram model (see Journel and Huijbregts (1978)). However, Germann et al. (2006) argued that using a simple variogram model or a covariance matrix may lead to unrealistic radar fields. As an alternative approach, copulas can be employed to describe the dependencies and to model multivariate random variables with different marginal distributions. In fact, describing the dependence structure independent of the marginal distribution is one of the most attractive features of copulas. The application of copulas in simulation of multivariate data and non-linear dependence structure has become popular in hydrological and meteorological studies. In recent years, a number of copulas-based models have been introduced for multivariate frequency analysis, risk assessment, geostatistical interpolation and multivariate extreme value analyses (see De Michele and Salvadori (2003); Bárdossy (2006); Genest et al. (2007); Renard and Lang (2007); Schölzel and Friederichs (2008); AghaKouchak et al. (2010a); Bárdossy and Li (2008); Zhang et al. (2008)). In the subsequent chapter a detailed discussion on the use of copulas in spatial modeling and rainfall simulation is provided.

Stochastically simulated rainfall ensembles have various applications in hydrological and meteorological applications. For example, using ensemble analysis one can evaluate flood prediction uncertainty and its associated risks for a given precipitation using an ensemble of precipitation estimates, instead of a single realization. Floods are the most life threatening hydrologic phenomena that result in extensive property damage and loss of lives each year. Ensemble analysis may improve assessment of flood forecast uncertainty which is influential in water resources management. Furthermore, climate change studies may also require simulated precipitation fields as possible future precipitation estimates in order to investigate long term changes in the hydrologic cycle of a specific region. Modeling climate changes with abnormal variations in one or more weather variables may allow exploration of subsequent effects on different hydrologic processes. For example, by projecting how precipitation may change over a significantly long period, one my be able to investigate future regional changes in surface runoff and subsurface flow. Evaluating these impacts is important yet challenging, as water budget, surface runoff and groundwater availability are sensitive to changes in weather variables, particularly precipitation. Applications of precipitation ensembles, are beyond the scope of this research; for additional information, interested readers are referred to Germann et al. (2009), Xuan et al. (2009) and Zhu et al. (2008).

1.2 OBJECTIVES

This study aims to investigate and develop different stochastic techniques for simulation of rainfall fields through simulating random error fields and imposing them over remotely sensed rainfall estimates. Four different models are developed and discussed in this study:

- In the first model, the Gaussian copula is employed for simulation of multivariate rainfall error fields. The outputs of the error simulator are then used to perturb radar based rainfall estimates.
- In the second model, another elliptical copula family, the t-copula, is used to describe the dependencies of rainfall error fields. Similar to the previous model, multiple synthetic rainfall realizations are obtained by imposing error estimates over radar estimates.
- In the third approach, a so-called v-transformed copula is applied for multivariate error simulations. Using this type of copula, one can describe asymmetrical dependencies through the copula parameters.
- In the last model, rainfall fields are generated by perturbing rainfall estimates with two normally distributed error terms: a purely random component and an error component that depends on rainfall rates.

In the first three models, having described the dependencies using copulas, the empirical distribution function of the observed rainfall error is numerically approximated and applied to the simulated error fields so that the simulated realizations are similar to those of the observed in terms of the distribution function. In the fourth model, however, the error is assumed to be normally distributed. In all cases, simulated error fields are imposed over remotely sensed rainfall estimates in order to obtain a rainfall ensemble.

Several case studies are performed over a small watershed in Mississippi, USA, and a large watershed in Oklahoma, USA to examine reliability, robustness and applicability of the models. Both radar reflectivity data (Level II) as well as Stage IV Next Generation Weather Radar (NEXRAD) multi-sensor precipitation estimates are used as input to the models. The NEXRAD system is installed and maintained by the National Weather Service (NWS). It consists of a network of Weather Surveillance Radars - 1988 Doppler (WSR-88D) that covers the entire United States (Young et al. (2000); Crum and Alberty (1993)). Using the presented models, ensembles of rainfall estimates are simulated by conditioning on available observations. The generated rainfall ensembles can then be used as input to hydrological and meteorological models to assess model prediction uncertainties. Subsequent runs of a hydrological or meteorological model using simulated realizations of radar fields would then allow an assessment of uncertainty propagation resulting from the precipitation input.

1.3 STRUCTURE OF THE STUDY

With respect to the scope of this study, the document is divided into 7 chapters. After the introduction, the study area and data resources are briefly introduced. In Chapter 3, an overview on copulas and the theoretical background is provided and the copula families used in this study are introduced. In the fourth chapter, two copula based models are developed and implemented for ensemble generation using two elliptical copula families (Gaussian copula and t-copula). In Chapter 5, a model based on a v-copula is introduced and applied for rainfall simulation. In the sixth chapter, a random error model is presented in which rainfall ensembles are simulated by imposing two normally distributed error terms, one purely random and one proportional to the magnitude of rain rates, over remotely sensed rainfall estimates. In Chapter 7, simulated ensembles obtained from the presented models are used as input into a physically based hydrological model to demonstrate implication of the ensemble generators in streamflow analysis. Chapter 8 is devoted to summary and conclusions and recommendations for further research.

CHAPTER 2

COPULAS

2.1 INTRODUCTION

Hydrological and meteorological phenomena are known to be multidimensional and thus require multivariate analyses as well as conditional probability distributions of variables (Genest and Favre (2007)). Classical families of multivariate distributions are commonly used for modeling joint probability distributions of several random variables. Classical multivariate distributions such as bivariate normal, log-normal and gamma are built with a number of model parameters that describe the behavior of each random variable as well as the joint probability distribution itself. The main disadvantage of such approaches is that modeling the dependence structure between variables is not independent of the choice of the marginal distributions (Genest and Favre (2007); Dupuis (2007)). The advent of copulas, however, allows hydrologists to avoid this restriction. The application of copulas in simulation of multivariate data, extreme value analysis and modeling dependence structure has become popular in hydrological analysis. In recent years, numerous copulas-based models have been introduced for different hydrologic applications. Kelly and Krzysoztofowicz (1997) employed a meta-Gaussian approach to bivariate rainfall analysis and pointed out that the meta-Gaussian distribution is independent of marginals and hence can be applied to represent the dependence structure of bivariate variables. De Michele and Salvadori (2003) used copulas to model intensity-duration of rainfall events. Favre et al. (2004) utilized copulas for multivariate hydrological frequency analysis. Bárdossy (2006) and Bárdossy and Li (2008) applied the concept of copulas to the interpolation of groundwater quality parameters. Zhang et al. (2008) carried out a bivariate rainfall frequency analysis using Archimedean copulas. Using the Gaussian copula, Renard and Lang (2007) investigated multivariate extreme value analysis and its application in hydrology. Schölzel and Friederichs (2008) modeled multivariate non-Gaussian random variables using copulas. Kuhn et al. (2007) employed copulas to describe spatial and temporal dependence of weekly precipitation extremes. Using bivariate mixed distributions (Shimizu (1993)) and a copula-based Markov approach, Serinaldi (2008), Serinaldi (2009) and Villarini et al. (2008) introduced a model for radar rainfall estimation uncertainties. A historical review and major developments in the theory and application of copulas can be found in Schweizer (1991), Kotz (1997) and DallÁglio (1991). In the subsequent sections, after a brief review on the theory of copulas, the copula families used in this study are introduced in more detail. Then, simulation using copulas is explained step by step.

2.2 THEORETICAL BACKGROUND

A copula C of n random variables is defined as a multivariate distribution function on the n-dimensional unit cube with uniform marginals:

$$C : [0,1]^n \rightarrow [0,1] \tag{2.1}$$

In other words, copulas are joint cumulative distribution functions that describe dependencies among variables independent of their marginals (Joe (1997); Nelsen (2006)):

$$C^n(u_1, \ldots, u_n) = Pr(U_1 \leq u_1, \ldots, U_n \leq u_n) \tag{2.2}$$

where C^n is an n-dimensional joint cumulative distribution function of a multivariate random vector $\boldsymbol{U}(U_1, \ldots, U_n)$ whose marginals are $u[0, 1]$. Note that throughout this document, a common statistical convention is used in which uppercase characters denote random variables and lowercase characters are their specified variables. Copulas are expected to satisfy the following conditions (Nelsen (2006); Malevergne and Sornette (2003)):

- $\forall u \in [0, 1], C(1, \ldots, 1, u, 1, \ldots, 1) = u$
- $\forall u_i \in [0, 1], C(u_1, \ldots, u_n) = 0$ if at least one of the u_i equals zero.
- C^n is grounded and n-increasing. This condition is satisfied if for all $u, \nu \in [0, 1]$ such that $u \leq \nu$, $V_c([u, \nu]) \geq 0$. The term $V_c([u, \nu])$ denotes the c-volume of $[u, \nu]$ (see Salvadori et al. (2007) for details).

In the theory of copulas, Sklar's theorem is fundamental to many applications. Sklar (1959) showed that each continuous multivariate distribution F can be represented with a unique copula C that can couple multivariate distribution functions to their corresponding marginal distribution functions:

$$F(x_1, \ldots, x_n) = C^n(F_1(x_1), \ldots, F_n(x_n)) \tag{2.3}$$

Note that the copula C^n is unique only if $F_1, \ldots, F_n$ are all continuous. Otherwise, the copula C^n is uniquely determined on $RanF_1, \ldots, RanF_n$, where Ran denotes range (Sklar (1996); Embrechts et al. (2001)). For proof and derivations, interested readers are referred to Sklar (1996). The Sklar theorem indicates that for multivariate distributions, the multivariate dependence structure and the univariate marginal distributions can be separated, and hence, the dependence structure can be represented by a copula independent of the marginals. Having described the dependencies using a copula, a transformation function can be applied to each variable in order to transform the marginal distribution into the desired marginals (Nelsen (2006)):

$$C^n(u_1, \ldots, u_n) = F(F_1^{-1}(u_1), \ldots, F_n^{-1}(u_n)) \tag{2.4}$$

where F is the multivariate cumulative distribution function (CDF) with marginals $F_1, \ldots, F_n$ belonging to different distribution families. In other words, using the Sklar theorem, one can simulate random variables with the same probability distribution as that of the input data while preserving the dependence structure of the variables. In the above Equation, for all x in $RanF$ the following relationship is valid: $F(F^{-1}(x)) = x$.

It is important to remark that copulas are invariant to monotonic transformations of the variables. That is, if $X_1, \ldots, X_n$ are continuous random variables with copula C^n and $F_1, \ldots, F_n$ are increasing functions on $RanX_1, \ldots, RanX_n$, then $F_1(X_1), \ldots, F_n(X_n)$ have the same copula C^n. Hence, the copula C^n is said to be invariant under monotonic transformations. In the following, the aforementioned property of copulas is explored in more detail as it is fundamental to the discussion in the following chapters. Assume $F_1, \ldots, F_n$ to be the distribution functions of $X_1, \ldots, X_n$, respectively. Consider monotonic transformations of the random variables $\Psi(X_1), \ldots, \Psi(X_n)$ with their corresponding marginals $G_1, \ldots, G_n$. Let C^n and C^n_Ψ be the copula of $X_1, \ldots, X_n$ and $\Psi(X_1), \ldots, \Psi(X_n)$, respectively. The strictly increasing property of Ψ indicates that for any $x \in R$ (Embrechts et al. (2001)):

$$G(x) = Pr\left(\Psi(X) \leq x\right) = Pr\left(X \leq \Psi^{-1}(x)\right) = F\left(\Psi^{-1}(x)\right) \tag{2.5}$$

and thus:

$$\begin{aligned} C^n_\Psi\left(G_1(x_1), \ldots, G_n(x_n)\right) &= Pr\left(\Psi_1(X_1) \leq x_1, \ldots, \Psi_n(X_n) \leq x_n\right) \\ &= Pr\left(X_1 \leq \Psi_1^{-1}(x_1), \ldots, X_n \leq \Psi_n^{-1}(x_n)\right) \\ &= C^n\left(F_1(\Psi_1^{-1}(x_1)), \ldots, F_n(\Psi_n^{-1}(x_n))\right) \\ &= C^n\left(G_1(x_1), \ldots, G_n(x_n)\right) \end{aligned} \tag{2.6}$$

Equation 2.6 confirms that $C^n = C^n_\Psi \in I^n$. This is a great advantage in simulation as the variables may belong to different probability distributions and applying transformation functions may be required to obtain the right marginals. It is noted that for a joint distribution function of $\boldsymbol{F}$, for example in Equation 2.3, the density function f is obtained by differentiating with respect to all variables (Melchiori (2003)):

$$f(x_1, \ldots, x_n) = c^n(F_1(x_1), \ldots, F_n(x_n))\Pi_{i=1}^n f_n(x_n) \tag{2.7}$$

where $f_1, \ldots, f_n$ are the density functions of the corresponding marginals $F_1, \ldots, F_n$ and $f(x_1, \ldots, x_n)$ is the density function of the joint distribution. c^n, which is termed as copula density, is basically the n-th partial derivative of an n-dimensional copula C^n (for derivations, the reader is pointed to Nelsen (2006) and Embrechts et al. (2001)):

$$c^n(u_1, \ldots, u_n) = \frac{\partial^n}{\partial u_1 \cdot \partial u_2 \ldots \partial u_n} C^n(u_1, \ldots, u_n) \tag{2.8}$$

There are many families of copulas developed for different practical contexts. Each family of copulas has a number of parameters to describe the dependencies. The main difference associated with different copulas is in the detail of the dependence they represent. For instance, various copula families may differ in the part of their distributions (upper tail/lower tail) where the association is strongest/weakest. In this study, two elliptical copulas, namely a normal copula and t-copula, as well as a non-Gaussian (v-transformed) copula are used for simulations. In the following, only the copula families employed in this work are discussed. For additional information regarding different copula families, the reader is referred to Nelsen (2006) and Joe (1997).

2.2.1 Gaussian Copula

The Gaussian copula, derived from the multivariate normal distribution, is perhaps the most commonly used copula family mainly due to its simplicity. The n-dimensional multivariate Gaussian copula with correlation matrix $\rho_{n\times n}$ can be expressed as (Nelsen (2006)):

$$C_\rho(u_1, ..., u_n) = F_\rho^n(F^{-1}(u_1), ..., F^{-1}(u_n)) \tag{2.9}$$

whose density function is:

$$c(u_1, \dots, u_n) = \frac{1}{\sqrt{det\rho}} exp\left(-\frac{1}{2}y(u)'(\rho^{-1} - I)y(u)\right) \tag{2.10}$$

where : F^n = Multivariate Gaussian CDF
$y(u_i)$ = $F^{-1}(u_i)$

2.2.2 t-Copula

The t-copula, also known as Student copula, is an elliptical copula based on the Student distribution that can be represented as:

$$C_{\nu,\rho}(u_1, \dots, u_n) = t_{\nu,\rho}^n(t_\nu^{-1}(u_1), \dots, t_\nu^{-1}(u_n)) \tag{2.11}$$

where : t^n = Multivariate Student CDF
ρ = shape matrix
ν = degrees of freedom

and

$$t^n_{\nu,\rho}(x) = \frac{1}{\sqrt{det\rho}} \frac{\Gamma\left(\frac{\nu+n}{2}\right)}{\Gamma\left(\frac{\nu}{2}\right)(\pi\nu)^{n/2}} \times \int_{-\infty}^{x_1} \cdots \int_{-\infty}^{x_n} \frac{dx}{\left(1+\frac{x'\rho^{-1}x}{\nu}\right)^{(\nu+n)/2}} \tag{2.12}$$

For $\nu > 2$, the shape matrix in Equation 2.11 is proportional to the correlation matrix (Malevergne and Sornette (2003)). Using Equation 2.8, the density function of the t-copula can be derived as (Malevergne and Sornette (2003)):

$$c(u_1,\ldots,u_n) = \frac{1}{\sqrt{det\rho}} \frac{\Gamma\left(\frac{\nu+n}{2}\right)\left(\Gamma\left(\frac{\nu}{2}\right)\right)^{n-1}}{\left(\Gamma\left(\frac{\nu+1}{2}\right)\right)^n} \times \frac{\Pi_{k=1}^n \left(1+\frac{y_k^2}{\nu}\right)^{(\nu+1)/2}}{\left(1+\frac{y'\rho^{-1}y}{\nu}\right)^{(\nu+n)/2}} \tag{2.13}$$

where : y_k = $t_\nu^{-1}(u_k)$
t_ν = univariate Student distribution with ν degrees of freedom

Figure 2.1 shows the bivariate copula density functions of the Gaussian copula and t-copula for different parameters. As shown, the density function is wider for lower correlations (compare Figures 2.1(a) and 2.1(b)). Additionally, note the difference between the density functions for different values of degrees of freedom (Figures 2.1(c) and 2.1(d)).

Both Gaussian and t-copulas are elliptical; however, they represent different tail dependencies, which describes the significance of the dependence in lower left quantile or upper right quantile of a multivariate distribution function. The upper tail expresses the probability occurrence of positive large values (outliers) at multiple locations jointly.

For a multivariate distribution with n random variables $X \equiv (X_1,\ldots,X_n)$, the upper tail dependence (λ_{up}) is described as (Melchiori (2003)):

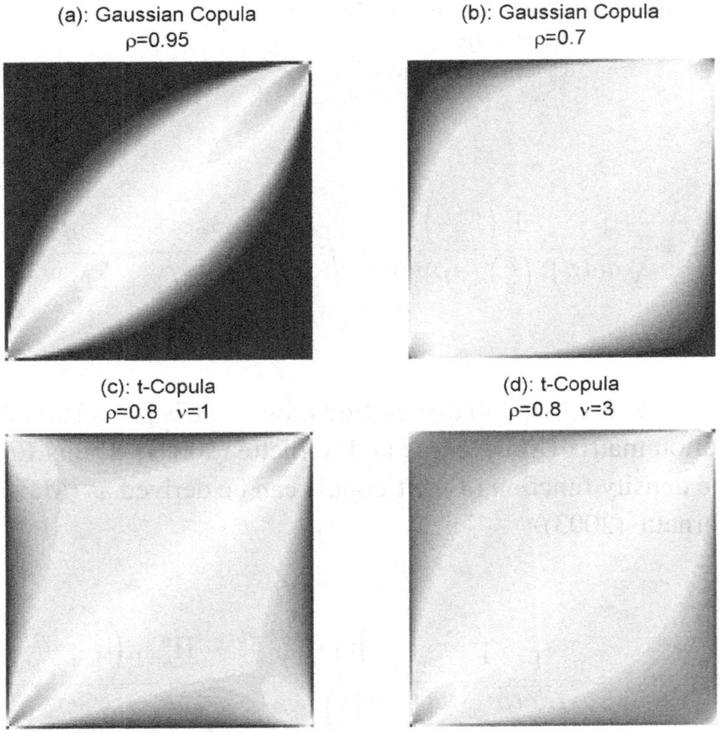

Figure 2.1 Bivariate copula density functions of the Gaussian copula and t-copula for different parameters.

$$\lambda_{up} = \lim_{u \to 1} Pr(X_1 \geq F_{X_1}^{-1}(u) | X_2 \geq F_{X_2}^{-1}(u) \ldots X_n \geq F_{X_n}^{-1}(u)) \tag{2.14}$$

where $F_1^{-1}, \ldots, F_n^{-1}$ are the inverse cumulative distributions of the random variables $X_1, \ldots, X_n$ and $u[0,1]$. In other words, Equation 2.14 indicates the probability occurrence of outliers in X_1, conditioned on presence of outliers in $X_2, \ldots, X_n$. The conditional probability given in Equation 2.14 can be expressed as:

$$Pr(X_1 \geq F_{X_1}^{-1}(u)|X_2 \geq F_{X_2}^{-1}(u) \ldots X_n \geq F_{X_n}^{-1}(u)) = \tag{2.15}$$

$$\frac{1 - Pr(X_1 \leq F_{X_1}^{-1}(u)) - \ldots - Pr(X_n \leq F_{X_n}^{-1}(u))}{1 - Pr(X_2 \leq F_{X_2}^{-1}(u)) - \ldots - Pr(X_n \leq F_{X_n}^{-1}(u))}$$
$$+ \frac{+Pr(X_1 \leq F_{X_1}^{-1}(u), \ldots, X_n \leq F_{X_n}^{-1}(u))}{1 - Pr(X_2 \leq F_{X_2}^{-1}(u)) - \ldots - Pr(X_n \leq F_{X_n}^{-1}(u))}$$

It is worth remembering that $Pr(X_1 \leq F_{X_1}^{-1}(u)) = \ldots = Pr(X_n \leq F_{X_n}^{-1}(u)) = u$ (Melchiori (2003)). Having Equation 2.2 in mind, the second term in the numerator of Equation 2.15 can be described as (Embrechts et al. (2001)):

$$C^{(n)}(u, \ldots, u) = Pr(X_1 \leq F_{X_1}^{-1}(u), \ldots, X_n \leq F_{X_n}^{-1}(u)) \tag{2.16}$$

where n is the dimension of the copula (number of random variables). Substituting Equation 2.16 into Equation 2.15 yields the following formulation for the upper tail:

$$\lambda_{up} = \lim_{u \to 1} \frac{1 - nu + C^{(n)}(u, \ldots, u)}{1 - (n-1)u} \tag{2.17}$$

Similarly, one can derive the lower tail ($\lambda_{lo} = \lim_{u \to 1} Pr(X_1 \leq F_{X_1}^{-1}(u)|X_2 \leq F_{X_2}^{-1}(u) \ldots X_n \leq F_{X_n}^{-1}(u))$) formulation as follows:

$$\lambda_{lo} = \lim_{u \to 1} \frac{C^{(n)}(u, \ldots, u)}{(n-1)u} \tag{2.18}$$

The joint distribution is said to be asymptotically dependent if $0 < \lambda_{up} \leq 1$ and asymptotically independent if $\lambda_{up} = 0$. The multivariate Gaussian copula shows asymptotic independence ($\lambda_{up} = 0$) regardless of the correlation between variables. That is, the extreme values in different variables occur independently even if there is a high correlation between variables. It is pointed out that for independent random variables, one could expect $\lambda_{up} = 0$; however, the converse does not hold (Malevergne and Sornette (2003)). That is, $\lambda_{up} = 0$ does not indicate that the variables are necessarily independent. While the Gaussian copula does not have upper or lower tail dependence, the t-copula can capture the upper tail dependence (if exists). The asymptotic dependent behavior of t-copula is expected even when the variables are negatively correlated (Embrechts et al. (2001)). As ν increases, the tail dependence becomes weaker and the probability occurrence of joint extremes reduces. Figure 2.2(a) illustrates the tail behavior of the bivariate t-copula with $\nu = 1$ to 10. The Figure shows occurrences of $x > 0.8$ (percentage) in both random vectors of the bivariate t-copula. It is noted that an increase in ν (degrees of freedom) results in less occurrences of extremes (values above a certain threshold). Figure 2.2(b), on the other hand, demonstrates the tail behavior of Gaussian copula. For a constant threshold of 0.8, occurrence of extremes in Gaussian copulas is significantly less than t-copula. This implies that for modeling the dependencies of extremes, the application of the Gaussian copula may not be an appropriate choice. It is worth pointing out that the tail behavior depends solely on the type of copula and not on the choice of marginal distribution. Thus, in copula-based simulation, the type of copula strongly affects the tail dependence of simulated realizations. For additional information on elliptical copulas, the reader is referred to Fang et al. (2002) and Genest et al. (2007)).

2.2.3 V-Transformed Copula

A v-transformed copula is obtained through a non-monotonic transformation of the multivariate Gaussian copula introduced in Equation 2.9 (Bárdossy and Li (2008)). The variable N is defined as an n-dimensional normal random variable with a mean of zero, unit standard deviation and correlation matrix ρ. The v-transformed copula can be derived with the following transformation (Bárdossy and Li (2008)):

$$X_i = \begin{cases} k(N_i - m) & \text{if } N_i \geq m \\ m - N_i & \text{if } N_i < m \end{cases} \tag{2.19}$$

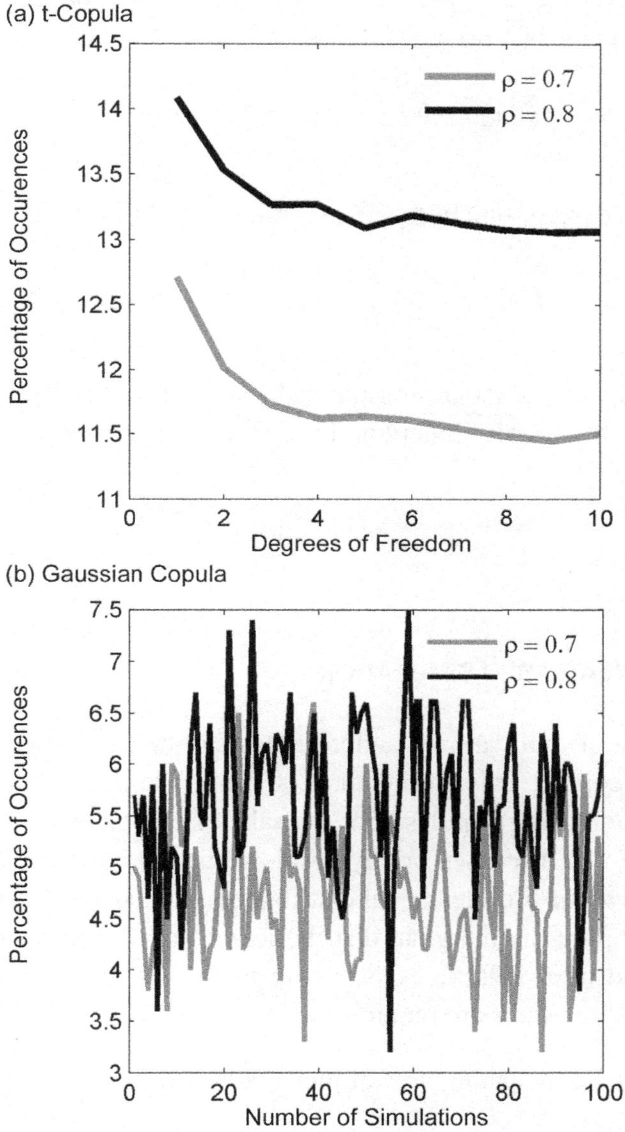

Figure 2.2 (a) Asymptotic dependent behavior of the t-copula: The tail dependence becomes weaker as ν increases; (b) Tail behavior of Gaussian copula.

where : k and m = copula parameters

The univariate marginal distribution of X_i can be expressed as (Bárdossy and Li (2008)):

$$F(x) = \Phi\left(\frac{x}{k} + m\right) - \Phi\left(-x + m\right) \quad (2.20)$$

with the density function being (Bárdossy and Li (2008)):

$$f(x) = \frac{1}{k}\varphi\left(\frac{x}{k} + m\right) + \varphi\left(-x + m\right) \quad (2.21)$$

where Φ is the standard normal distribution and the term φ denotes the probability density function of the normal distribution:

$$\varphi = \frac{1}{\sqrt{2\pi\sigma^2}} EXP\left(-\frac{(x-\mu)^2}{2\sigma^2}\right) \quad (2.22)$$

2.3 COPULA-BASED SIMULATION

Simulation of multivariate random fields can be performed based on any given copula. Salvadori et al. (2007) provides a general overview of simulation using copulas. For the sake of completeness, the simulation steps are discussed briefly in this section. Let C^n be the copula of a multivariate n-dimensional distribution $H \equiv (H_1, \ldots, H_n)$ where $H_1, \ldots, H_n$ are the marginal distributions. In order to obtain a simulated field of $x \equiv (x_1, \ldots, x_n)$ with marginals of $H_1, \ldots, H_n$, the following three steps are required:

1. Estimate the parameters of the copula C^n.

2. Simulate uniform random variables $\boldsymbol{u}(u_1, \ldots, u_n)$ using the copula C^n.

3. Transform the univariate marginals to $H_1, \ldots, H_n$ using Sklar's theorem (Sklar (1959)): $x_i = H_i^{-1}(u_i)$.

It is noted that $H_1, \ldots, H_n$ do not necessarily need to belong the same distribution family nor are they restricted to standard probability distributions. Furthermore, since copulas are invariant to transforms, the simulated random variable $\boldsymbol{x}$ will have the same spatial dependence structure as that of $\boldsymbol{u}$. This fact is one of the main attractive features of using copulas for the simulation of spatially dependent random fields. Performing the above three steps results in a simulated field of dependent random variables $x_1, \ldots, x_n$ with the marginal distributions of $H_1, \ldots, H_n$. While the first and third steps are straightforward, the second step requires additional attention. Let $u_1, \ldots, u_n$ be simulated random vectors with marginal distributions of $U_1, \ldots, U_n \in [0, 1]$. The first random variable, for example u_i, can be sampled from $\acute{U}_i \in [0, 1]$. Then, in order to simulate u_i and u_j $(i, j = 1, \ldots, n)$ based on the joint distribution function $\boldsymbol{F}$, the distribution of u_j must be conditioned on u_i as $u_j = F_j^{-1}(\acute{u}_j|u_i)$ where $\acute{u}_j$ is a random realization of $\acute{U}_j$ independent of $\acute{U}_i$. In an n-dimensional case, the latter equation can be expressed as (after Salvadori et al. (2007)):

$$u_n = F_n^{-1}(\acute{u}_n|u_1, \ldots, u_{n-1}) \tag{2.23}$$

It is worth remembering that in terms of probabilities, the simulation of a random variable u_n conditioned on $n - 1$ other random variables can be described as (Salvadori et al. (2007):

$$F_n(u_n|u_1, \ldots, u_{n-1}) = Pr(U_n \leq u_n|U_1 = u_1, \ldots, U_{n-1} = u_{n-1}) \tag{2.24}$$

Considering Equation 2.2 and Equation 2.24, the general formulation of multivariate simulation using a conditional copula can be expressed as:

$$F_n(u_n|u_1, \ldots, u_{n-1}) \approx \frac{\partial_{u_1,\ldots,u_{n-1}} C(u_1, \ldots, u_k)}{\partial_{u_1,\ldots,u_{n-1}} C(u_1, \ldots, u_{k-1})} \tag{2.25}$$

When conditionally simulating multivariate random fields, the aim is to generate multiple random variables at a location x where no observation is available on the basis of n observed values. It should be noted that the simulation method must honor observed values at their locations. In order to simulate the random variable y at location x based on $y_1, \ldots, y_n$, the following conditional copula C_x^n is required:

$$F_{(x,y)}^n = C_x^n \left(F(y) | u_1 = F(y_1), \ldots, u_n = F(y_n) \right) \tag{2.26}$$

where n is the number of available observations. Note that for large values of n, the above derivation may become too computationally expensive. Thus, for large values of n, one may opt to only include the observations in the neighborhood of x. It is also noted that the the conditional copula density c_x^n for variable z at location x can be approximated as (Bárdossy (2006)):

$$c_x^n \left(u_1 | u_2 = F(z_2), \ldots, u_n = F(z_n) \right) \approx \frac{f_n(z_1, \ldots, z_n)}{f_1(z_1) \cdot f_1(z_2) \ldots f_1(z_n)} \tag{2.27}$$

where the terms $F(z_i)$ and $f(z_i)$ refer to the corresponding distribution and density functions of the multivariate joint distribution. For the v-transformed copula, for example, f_1 can be obtained from Equation 2.21. Using the above formulation, multivariate random fields with a given correlation structure can be generated (Step 2). Having simulated multivariate uniform random variables based on copula C, one can use the probability integral transform, explained in step 3, to apply the right marginal distributions.

CHAPTER 3

STUDY AREAS AND DATA RESOURCES

The presented models were performed over a small watershed in Mississippi, USA, and a large watershed in Oklahoma, USA to examine model reliability, robustness and applicability. A brief overview of both watersheds is given in Table 3.1. Additional details about the watersheds and the data used in this study are presented in the following sections.

3.1 THE LITTLE WASHITA WATERSHED

The Little Washita watershed, located in the southwestern part of the state of Oklahoma is one of the largest and most highly instrumented watersheds in the United States and has been the target of remote sensing, rainfall analysis and climate change research over the past few decades (e.g. Ciach et al. (2003), Young et al. (2000), Habib et al.

(2004)). The area of the watershed is approximately 611 square kilometers and the surface water drains into the Little Washita River which is situated between Chickasha and Lawton and is a tributary of the Washita River. Hydrological and meteorological variables of the watershed have been measured in a variety of hydrologic research projects since the 1930's. The climate of the region is classified as moist subhumid and the average annual rainfall of the watershed is approximately 760 mm, and the mean annual temperature has been determined to be 16 degrees Celsius (Elliott et al. (1993)). While summers are hot and long with an average daily high temperature of 34 degrees Celsius for July, winters are relatively short and temperate with the average daily low temperature of -4 degrees Celsius for January. Most major rainfall events, and majority of the annual precipitation, occur in the spring and fall. A more detailed description of the climate variables of the Little Washita watershed is provided in Staff and Laboratory (1983). The Agricultural Research Service (ARS) started collecting and managing hydrometeorological data on the Little Washita since the 1960's. Thirty years later, the watershed became one of the main study sites for a variety of ARS research projects dealing with water resources and climate change.

The Micronet network, operated by Agricultural Research Service (ARS), is located within and around the Little Washita boundaries. The Micronet network includes 42 stations, almost uniformly distributed throughout the watershed. The data includes rainfall accumulations, temperature, relative humidity, solar radiation, etc. In addition to the Micronet network, there are three stations of Oklahoma Mesonet within the study area that are not included in the analysis, as they are used in the multi-sensor precipitation estimation process. The Micronet network is equipped with tipping-bucket gauges that record rainfall data with a temporal resolution of 5 minutes and accuracy of 0.254 mm (Young et al. (2000)). The 5 min accumulations are aggregated to hourly intervals in order to synchronize the rain gauge measurements with radar rainfall estimates. The reference surface rainfall data are then used to obtain estimates of the radar rainfall error (observed error) across the study area. The observed error estimates are used for simulation of random error fields as explained in subsequent chapters. Figure 3.1 shows the location of the Little Washita watershed and the position of rain gauges (marked with circle symbols) and radar pixels (4 km × 4 km grids).

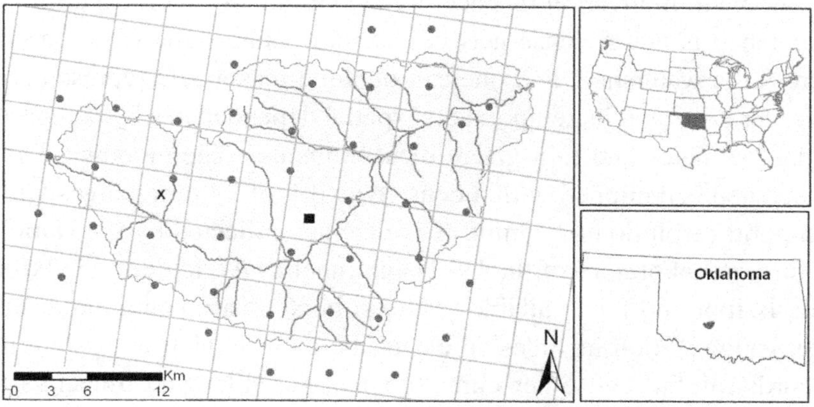

Figure 3.1 Little Washita research watershed, Oklahoma, USA.

3.2 THE GOODWIN CREEK WATERSHED

The Goodwin Creek experimental watershed is located in the north central part of the state of Mississippi near Batesville, USA. The area of the watershed is approximately 21 km^2 with the outlet at latitude 34°13'55" N and longitude 89°54'50" W. The watershed is situated within the Yazoo River basin and is instrumented for conducting extensive research on watershed hydrology. The Goodwin Creek drains into the Yocona River, a tributary of the Yazoo River, which flows into the Mississippi River. Its average annual rainfall, measured at the climatological station near the center of the watershed, is 1440 mm, and the mean annual runoff is approximately 144 mm at the watershed outlet. Much of the annual rainfall and occurs in the spring and fall. The average daily maximum temperatures were reported as 30 °C and 10 °C in summer and winter, respectively. The topographic elevation ranges from 71 m at the outlet to 128 m at the watershed divide with reference to mean sea level and there is an average channel slope of 0.004. A more detailed review of the climate variables of the Goodwin Creek watershed is given in Alonso and Binger (2000). Table 3.1 compares the Little Washita and Goodwin Creek watersheds.

The main motivation in selecting the Goodwin Creek experimental watershed is that a dense network of rain gauge stations is available across the watershed, and one can obtain a reasonable representation of radar error and its temporal and spatial dependencies by comparing radar estimates and rain gauge measurements. The Goodwin Creek watershed is equipped with a dense network of 29 rain gauges with a temporal resolution of 15min. The data are collected by the National Sediment Laboratory of the US Department of Agriculture in Oxford, Mississippi and are available to the public. Instrumentation at each rain gauge station includes an electronic data acquisition system that records rainfall and other climatological variables and transmits the measurements to the National Sediment Laboratory. Figure 3.2 shows the position of rain gauges (marked with circle symbols) and radar pixels (1 km^2 grids) across the watershed.

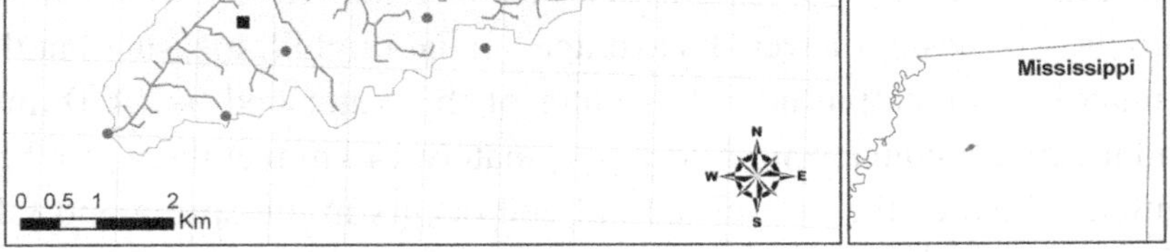

Figure 3.2 Goodwin Creek experimental watershed, Mississippi, USA.

Table 3.1 Comparison of the Little Washita and Goodwin Creek watersheds.

	Little Washita	Goodwin Creek
Location	Oklahoma, USA	Mississippi, USA
Area (km^2)	611	21.3
Average annual rainfall (mm)	760	1440
Average annual temperature ($^\circ$)	16	18
Distance from radar (km)	70	110
Climate zone	sub-humid	humid

3.3 DATA RESOURCES

For the Goodwin Creek watershed, the NEXRAD Level II reflectivity (Z) radar data of the Memphis radar site, recorded at the National Weather Service (NWS) is used for analysis (Crum and Alberty (1993)). The radar station is a 1988 Doppler (WSR-88D), located approximately 110 km from the center of the watershed. Radar reflectivity data are available in polar coordinates at a resolution of 1° for different elevation angles starting at 0.5° and with a time sampling interval of 5 to 6 minutes. It is noted that the temporal resolution of the Goodwin Creek gauge data is 15 minutes, and thus, the radar rainfall data was aggregated into 15 min intervals in order to synchronize radar estimates with rain gauge measurements. Since the interest is in the surface rainfall information, a regular 1 km $\times$ 1 km Cartesian reflectivity grid is obtained from the lowest elevation angle of the polar reflectivity beam.

The relationship between radar reflectivity factor Z and rain rate R is approximated by a power law function of the form $(Z/A)^{1/b}$ where A and b are the multiplicative factor and exponent of the $Z-R$ relationship, respectively. Before, using Level II reflectivity data, the $Z-R$ relationship parameters (A and b) are to be estimated first. In this study, the parameters A and b are estimated based on Steiner and

Smith (2000) whereby for a range of b values, A is estimated using Equation 3.1 (Steiner and Smith (2000)):

$$A = \left(\frac{\sum_{i=1}^{m} (Z_i)^{1/b}}{\sum_{i=1}^{m} R_i} \right)^b \tag{3.1}$$

In Equation 3.1, Z_i and R_i are pair of reflectivity and ground reference measurements used for parameter estimation. For the estimated A and b values, the Root Mean Square Error ($RMSE$) is calculated using Equation 3.2. Then, the combination of A and b values that leads to the least error are taken as the estimated parameter.

$$RMSE = \frac{\sqrt{\frac{1}{m-2} \sum_{i=1}^{m} \left(R_i - (Z_i/A)^{1/b}\right)^2}}{\frac{1}{m} \sum_{i=1}^{m} R_i} \tag{3.2}$$

It is worth pointing out that by using this method the overall bias between radar estimates and rain gauge measurements is removed (Steiner and Smith (2000)). That is, the total amount of radar estimates and gauge measurements over a given storm will be the same as A is estimated for each b such that the overall bias is removed. Previous studies show that the multiplicative factor exhibits more variability than the exponent from event to event (Steiner and Smith (2000)). This considerable variability in the multiplicative factor may result in significant uncertainty in the $Z - R$ relationship. Hence, in this study, the multiplicative factor and exponent are adjusted for each event separately as recommended by (Steiner and Smith (2000)). Note that further adjustments within a given storm (e.g. different parameters for stratiform and convective rainfall events) may be desired but are typically not practical (Steiner and Smith (2000)).

For the Little Washita watershed, the Stage IV radar-based multi-sensor precipitation estimates (MPE) provided by NCAR/EOL, under sponsorship of the National Science Foundation (NSF), are used for analysis. The Stage IV data are generated in near real-time by mosaicking the multi-sensor precipitation estimates from the National Weather Service (NWS) River Forecast Centers (RFC). Based on the radar data collected over the entire Continental United States, the Stage IV data are provided with a spatial resolution of 4 km $\times$ 4 km termed Hydrologic Rainfall Analysis Project (HRAP) which is a national grid system where each pixel corresponds to an MPE value. The MPE data are obtained from merging operational radar estimates and automated rain gauge measurements following quality control and bias adjustment. The data are available with a temporal resolution of 1 hour. The Stage IV data are then accumulated from hourly into 3-hour, 6-hour, daily and monthly temporal resolutions for further analysis. The MPE data are now widely used in hydrologic forecast models, decision making, flood and flash flood warning and other hydrologic monitoring purposes across the United States.

It is noted that Stage IV data are calibrated rainfall estimates, available in rain rates. Therefore, unlike Level II reflectivity data, no transformation function is required to obtain the actual rain rates. However, in order to convert the Stage IV data to a format suitable for use in image processing, the following post-processing methods are necessary: data acquisition; data conversion and georeferencing; conversion to Raster data; data export. In these steps, Stage IV estimates are converted from the radial GRIB format into raster based data. In this study, the Stage IV data with an hourly temporal resolution is used for analysis. The closest radar station to the study area is the Oklahoma City operational radar site which is located approximately 70 km from the center of the watershed.

3.4 RAINFALL EVENTS

In order to evaluate the performance of the presented models, ensembles of rainfall fields are simulated for a number of rainfall events over different temporal and spatial scales. The rainfall events that occurred over the Little Washita watershed are identified with LW, whereas the events that occurred across the Goodwin Creek watershed are distinguished with GC. The notations, II and IV (in GC1-II, LW1-IV) refer to the Level II reflectivity and Stage IV radar data, respectively. Stage IV radar data, with the finest available temporal and spatial resolutions

(1 hr, 4 km $\times$ 4 km) are processed for simulations. The Level II reflectivity data are used with a spatial resolution of 1 km^2 and temporal resolutions of 15 min. Table 3.2 displays summary statistics including mean, standard deviation, 10 and 90 percent quantiles of the lumped rainfall accumulation for each storm. Note that for the Goodwin Creek Level II reflectivity rainfall data, the parameters of the $Z - R$ relationship (A and b) are estimated using the method explained in Section 3.3 as listed in Table 3.3. The estimated parameters fit well within the wide range of values reported in the literature (Doelling et al. (1998); Smith and Krajewski (1993); Battan (1973); Stout and Mueller (1968); Ulbrich (1978)).

Table 3.2 Summary statistics of rainfall accumulations for the selected rainfall events.

Event ID	Date Time	Duration [h]	Mean [mm]	Standard Deviation	Q10 [mm]	Q90 [mm]
LW1-IV	8/30/03 07:00	12:00	4.3	5.5	0.8	8.8
LW2-IV	9/15/05 06:00	14:00	3.8	4.3	1.0	10.7
LW3-IV	9/17/06 15:00	10:00	10.1	15.5	1.0	22.7
LW4-IV	8/18/07 21:00	27:00	9.3	13.5	1.5	27.1
LW5-IV	8/18/08 10:00	50:00	4.1	5.9	1.4	7.1
GC1-II	1/24/02 03:30	16:15	10.4	10.4	1.1	19.3
GC2-II	3/20/02 01:45	19:00	6.9	14.2	1.5	10.6
GC3-II	3/29/02 23:15	32:15	8.5	15.6	1.1	9.2
GC4-II	5/02/02 13:45	34:25	9.1	21.9	1.1	21.4
GC5-II	9/25/02 06:30	31:30	12.3	19.4	1.9	10.3

Table 3.3 Estimated parameters (A and b) of the $Z - R$ relationship.

Event ID	A	b	$Z - R$
GC1-II	203.1	1.50	$Z = 203.1R^{1.5}$
GC2-II	235.8	1.45	$Z = 235.8R^{1.45}$
GC3-II	205.1	1.45	$Z = 205.1R^{1.45}$
GC4-II	223.9	1.35	$Z = 223.9R^{1.35}$
GC5-II	117.6	1.35	$Z = 117.6R^{1.35}$

CHAPTER 4

SIMULATION USING ELLIPTICAL COPULAS

4.1 INTRODUCTION

As mentioned in Chapter 1, previous studies confirm that efforts are required to determine the accuracy of remotely sensed rainfall estimates and their associated uncertainties, as these values are of great significance to the continuing research and applications in hydrology and climate studies (Habib et al. (2008); Ciach et al. (2007); Villarini et al. (2009); Margulis et al. (2006); Dong et al. (2005); Hossain and Huffman (2008); Lucieer and Kraak (2004); Crosetto et al. (2001); Bastin et al. (2002)). In recent years, various approaches and extensive research have been undertaken to develop an uncertainty model for remotely sensed rainfall estimates (see Mehrotra et al. (2006) and references therein). In this chapter two ensemble generators are developed and presented for radar reflectivity (Level II) as well as Stage

IV radar-based multi-sensor precipitation estimates (MPE) that can be used to evaluate the uncertainty of remotely sensed rainfall estimates. The Gaussian copula forms the basis of the first model, whereas the second model employs the t-copula for conditional simulations of rainfall error (uncertainty) fields. Simulated fields are then imposed on radar estimates to obtain an ensemble of precipitation estimates that can be used for uncertainty assessment. The models are implemented for various rainfall events over the study areas introduced in Chapter 3. The simulated rainfall fields are then investigated with respect to statistical properties, extreme values and spatio-temporal dependencies.

4.2 MODEL DESCRIPTION

One way to account for uncertainties associated with remotely sensed rainfall estimates is to generate an ensemble of rainfall realizations, that is a representation of possible variabilities in precipitation data. Multiple realizations of rainfall fields can be obtained by imposing random error fields over the original rainfall estimates (e.g. radar rainfall images). In the copula-based approach implemented here, multiple error fields (ϵ) are simulated using the Gaussian copula and t-copula and imposed on radar rainfall fields (R_i). Equation 4.1 describes the general formulation of the model:

$$R_s = R_i + R_i \times \epsilon \tag{4.1}$$

where : R_s = simulated rainfall fields
ϵ = uncertainty component

Previous studies indicate that radar rainfall error is proportional to the magnitude of rain rate (Ciach et al. (2007); Habib et al. (2008); Villarini et al. (2009); Habib et al. (2009)). The importance of error simulation conditioned on the magnitude of rain rate has been addressed in numerous studies (e.g. Habib et al. (2008), AghaKouchak et al. (2010b) and references therein). Intuitively, one can argue that while large rain rates are subject to large or small random error, small rainfall estimates are not subject to very large random error. This characteristics of rainfall error is included in the second term of Equation

4.1, where rainfall uncertainty is multiplied by the magnitude of rainfall rate ($R_i \times \epsilon$). Using this formulation, the simulated error fields ($R_i \times \epsilon$) will be proportional to the magnitude of rain rate. Application of a multiplicative error component for uncertainty of rating curves has been tested and verified in previous studies (e.g. Petersen-Ø verleir (2004) and Seber and Wild (1998)). For example, Petersen-Ø verleir (2004) shows that a multiplicative error term can be used to describe the uncertainty of measured rating curves. This approach is adopted here to avoid unrealistically large simulated errors for insignificant rain rates.

Previous research on properties of radar rainfall error shows that the spatial correlation of error is not negligible (Habib et al. (2008); Ciach et al. (2007)). Therefore, the error term (ϵ) in Equation 4.1 is assumed to be spatially correlated and copulas are used to describe the dependence structure. In order to obtain estimates of the radar rainfall error across the study area, reference surface rainfall data are obtained from high resolution rain gauge measurements over the study areas. Further details on the spatial and temporal dependencies of rainfall error are presented in Section 4.5.

4.3 PARAMETER ESTIMATION

In order to simulate a random field using copulas, the copula parameters are to be estimated first. The first model introduced here is based on the multivariate Gaussian copula which is identified through a correlation matrix $\rho_{n\times n}$ where n is the dimension of the copula. It is worth remembering that the correlation matrix $\rho_{n\times n}$ is an $n \times n$ symmetrical matrix that can be expressed as a parameter dependent on different distance vectors of d: $\rho_{ij} = \xi(x_i - x_j)$ where the function $\xi(d)$ denotes a positive-definite correlation function such as a correlation function corresponding to the spherical variogram (Bárdossy and Li (2008)). This method is employed to estimate the correlation matrix ($\rho_{n\times n}$) of the multivariate Gaussian copula. Assuming a spherical function for $\xi(d)$, the parameters of the function are estimated as described in Bárdossy and Li (2008). The function is then used to obtain the correlation matrix of the multivariate Gaussian copula. In this study, for each event, the parameters are estimated based on the seasonal observations (e.g. summer 2003, summer 2005) in which the event occurs. The reason for this is explained further below. Table 4.1 gives the parameters of the spherical function used in this study. The parameters s_1, s_2, r_1 and r_2 represent the sills and ranges of a function consisting of two spherical components. The parameter s_0, denotes the sill of a nugget effect in the aforementioned function.

Table 4.1 The parameters of the spherical function.

Event ID	s_0	s_1	s_2	r_1 [m]	r_2 [m]
LW1-IV	0.04	0.11	0.31	3100	14500
LW2-IV	0.06	0.13	0.39	4000	13000
LW3-IV	0.01	0.19	0.41	4300	14500
LW4-IV	0.02	0.22	0.52	3700	17500
LW5-IV	0.09	0.26	0.33	4400	12000
GC1-II	0.07	0.29	0.29	3600	7500
GC2-II	0.01	0.25	0.42	2200	6000
GC3-II	0.01	0.25	0.42	2200	6000
GC4-II	0.01	0.25	0.42	2200	6000
GC5-II	0.05	0.11	0.38	2400	7000

The parameters of the t-copula are the correlation matrix and degrees of freedom ν (Equation 2.11). For the t-copula, the degrees of freedom is estimated using the Inference Function for Margins (IFM) method (Joe (1997); McLeish and Small (1988)) which is a special case of the Generalized Method of Moments (GMM) with an identity weight matrix (Davidson and MascKinnon (1993)). The IFM methods can be used for estimation of univariate as well as dependence parameters from separate univariate and multivariate likelihood functions (Purwono (2005)). In this approach, the model parameters are estimated via a nonlinear system of equations, each of which being a partial derivative (score function) of a log-likelihood function from some marginal distribution of the multivariate model (Joe and Xu (1996)). In the following the IFM method is briefly discussed.

Having Equation 2.3 and its derivative (Equation 2.7) in mind, the loglikeihood function for an n-dimensional copula can be expressed as:

$$L_{\nu,\rho}(x_1,\ldots,x_n) = \Sigma_{i=1}^{n} \log f_{\nu,\rho}((x_1,\ldots,x_n) \tag{4.2}$$

By taking the partial derivatives of the above equation with respect to $\nu_1,\ldots,\nu_n$, the maximum likelihood estimator (MLE), also known as the set of inference functions, for the parameters can be derived, as shown in Equation 4.3, for an n-dimensional system (Purwono (2005)):

$$\left(\frac{L^1_{\nu,\rho}(x_1,\ldots,x_n)}{\partial\nu_1},\ldots,\frac{L^n_{\nu,\rho}(x_1,\ldots,x_n)}{\partial\nu_n}\right) = 0 \tag{4.3}$$

The parameters $\nu_1,\ldots,\nu_n$ are then estimated by finding the roots of the defined sets of inference functions in Equation 4.3. By substituting $\nu_1,\ldots,\nu_n$ into Equation 4.2 and taking the partial derivatives with respect to the dependence parameter, one can obtain the following:

$$\left(\frac{L^1_{\nu,\rho}(x_1,\ldots,x_n)}{\partial\rho_{1,1}},\ldots,\frac{L^n_{\nu,\rho}(x_1,\ldots,x_n)}{\partial\rho_{n,n}}\right) = 0 \tag{4.4}$$

Using the above set of equations, the dependence parameter can be estimated numerically. Joe (1997) argues that the IFM method is highly efficient and reliable for parameter estimation and that this approach is computationally less extensive than using optimization techniques for parameter estimation. However, some statistical tests are required to make sure that the estimated parameters are reliable and that the available data is sufficient for parameter estimation. For this reason, the parameters are estimated based on four different types of data:

- Case 1: for each rainfall event separately.

- Case 2: seasonal data that each rainfall event occurred in (e.g. for Event LW1-IV, the entire summer 2003)
- Case 3: random subsets of the available data for each rainfall event (subsets are obtained from Case 1).
- Case 4: random subsets of the seasonal data for each event (subsets are obtained from Case 2).

Ideally, one may opt to estimate the parameters based on the available data for each rainfall event of interest separately. However, the data may not be sufficient for parameter estimation. Therefore, parameter estimation based on seasonal data is adopted to investigate this issue. In this approach, instead of using the data of one event, the entire season in which the event occurs will be used for parameter estimation. For example, in Event LW1-IV which occurs in summer 2003, the entire summer (June, July and August) data is used for parameter estimation. The mean and confidence interval of 100 randomly selected subsets of the data in Case 1 and Case 2 are also calculated and compared to the results obtained from Case 1 and 2. Figure 4.1 plots the estimated degrees of freedom for each rainfall event using event-based (Case 1), and seasonal data (Case 2). Additionally, the graph shows the mean and confidence interval ($\pm$ 3 times the standard error) of the estimated degrees of freedom for each rainfall event using random subsets of the data (Case 3 and Case 4). The solid black lines represent the confidence intervals of the estimated parameters based on the random subsets of the seasonal data (Case 4), whereas the gray lines show the confidence intervals of the estimated parameters based on random subsets of the available data for each event (Case 3). Here, the standard error is used to provide an indication of amount of uncertainty. It is noted, however, that the estimation of the confidence intervals using the standard error is only acceptable when the sample size is *large* or at least *moderately large* (Bernardo and Smith (2000); Cox and Hinkley (1974)).

As seen in Figure 4.1, the estimated degrees of freedom (ν) based on the seasonal data (Case 2) fall within the confidence intervals of the estimated ν from the corresponding random subsets (Case 4). However,

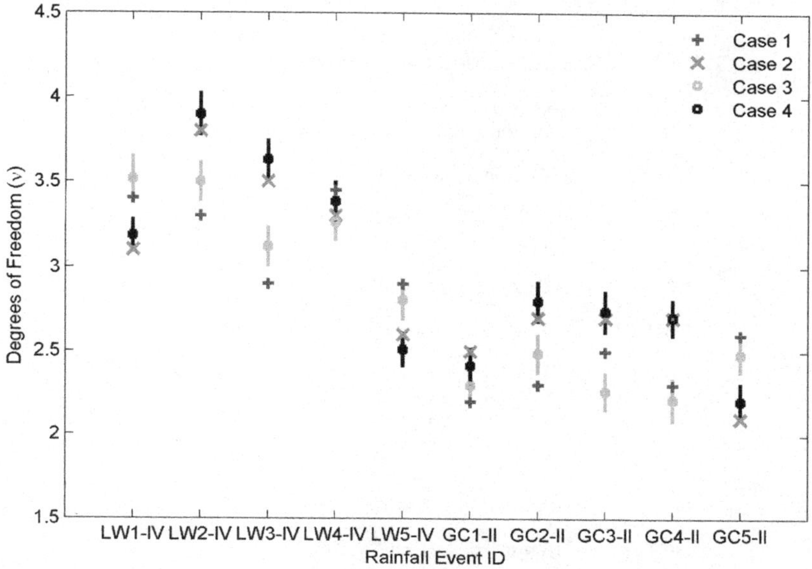

Figure 4.1 The estimated degrees of freedom for each rainfall event using event-based (Case 1), and seasonal data (Case 2). The solid black lines represents the confidence intervals of the random subsets of the seasonal data (Case 4) whereas, the gray lines show the confidence intervals of the random subsets of the available data for each event (Case 3).

in some rainfall events (e.g. LW2-IV, LW3-IV, GC2-II, GC3-II), the event-based estimates (Case 1) did not fall within the range of the estimated ν from the random subsets. This may be due to the fact that the available data in each event were not sufficient to estimate the parameters reliably (Bouye et al. (2000); Purwono (2005)). Therefore, the value of the degrees of freedom for each rainfall event is estimated based on the seasonal data as listed in Table 4.2. In subsequent sections, the fitted copula is tested through numerous statistical tests.

4.4 RAINFALL SIMULATION

Having estimated the copula parameters as explained in Section 4.3, multiple univariate random fields are simulated using the Gaussian copula and t-copula (described in Chapter 2). The marginals of the observed error are then applied to the univariate random fields using the

Table 4.2 The estimated degrees of freedom for the rainfall events.

Event ID	Temporal Resolution	Spatial Resolution	Degree of Freedom ν
LW1-IV	1 hr	16 km^2	3.1
LW2-IV	1 hr	16 km^2	3.8
LW3-IV	1 hr	16 km^2	3.5
LW4-IV	1 hr	16 km^2	3.3
LW5-IV	1 hr	16 km^2	2.6
GC1-II	15 min	1 km^2	2.5
GC2-II	15 min	1 km^2	2.7
GC3-II	15 min	1 km^2	2.7
GC4-II	15 min	1 km^2	2.7
GC5-II	15 min	1 km^2	2.1

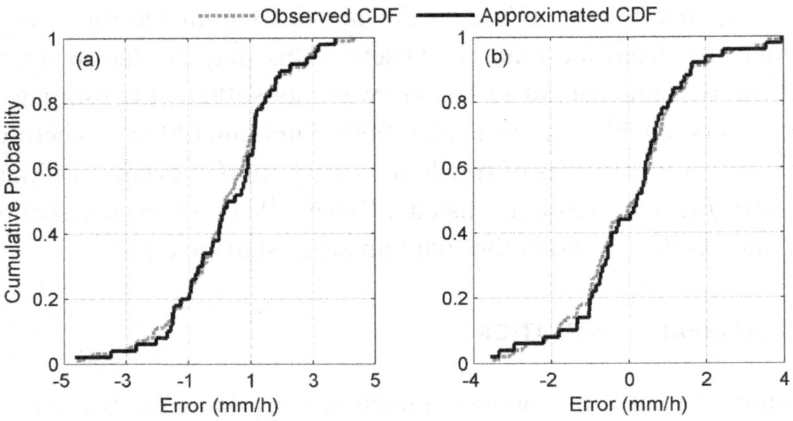

Figure 4.2 Numerical approximation of inverse CDFs using Stair functions.

Sklar Theorem (Equation 2.3). In this work, the empirical cumulative distribution function (CDF) of the observed error is numerically approximated using a Stair function and applied on the simulated uniform fields. The empirical inverse CDF of the observed data is basically a stair function, with steps at the values $1/nb, 2/nb, \ldots, 1$ where nb is the number of bins and the data in each bin is the step height. Figures 4.2(a) and 4.2(b) show two numerically approximated inverse CDF functions of rainfall error over the square-marked pixel shown in Figure 3.1 during Events LW1-IV and LW2-IV. The dotted lines represent the empirical CDFs of the observed error whereas, the solid lines show the numerically approximated empirical CDFs. It is worth remembering that copulas are invariant to monotonic transformations, and hence the simulated error fields will have the same spatial dependence structure after the transformation.

As discussed in section 4.2, in both models (Gaussian copula and t-copula) multiple random error fields are simulated using a multiplicative error component and imposed over radar-based rainfall estimates in order to obtain an ensemble of rainfall realizations. Figure 4.3(a) presents a radar-based rainfall estimates occurred during Event LW1-IV and Figure 4.3(b) shows the corresponding rain gauge measurements. Figures 4.3(c) and 4.3(d) display two realizations of conditionally simulated rainfall fields using Gaussian copula and t-copula, respectively. Similar figures are presented for Events LW2-IV to LW5-IV (Figures 4.3(e) to 4.3(t)). A visual comparison confirms that both models reproduce the rain gauge measurements (Figures 4.3(b), 4.3(f), 4.3(j), 4.3(n) and 4.3(r)) at their locations (see Figures 4.3(c), 4.3(d), 4.3(g), 4.3(h), 4.3(k), 4.3(l), 4.3(o), 4.3(p), 4.3(s), 4.3(t)). Note that in the first (LW1-IV) and second (LW2-IV) events, 42 and 22 rain gauges were available, respectively, whereas in the other events (LW3-IV, LW4-IV and LW5-IV), 20 rain gauges were in operation and available for analysis. Similar figures for the rainfall events GC1-IV to GC5-IV are provided in Figures 4.4(a) to 4.4(t). The third column shows simulated fields using the Gaussian copula, while the fourth column (Figures 4.4(b), 4.4(d), 4.4(f), 4.4(h) and 4.4(j)) presents rainfall ensembles obtained using the t-copula.

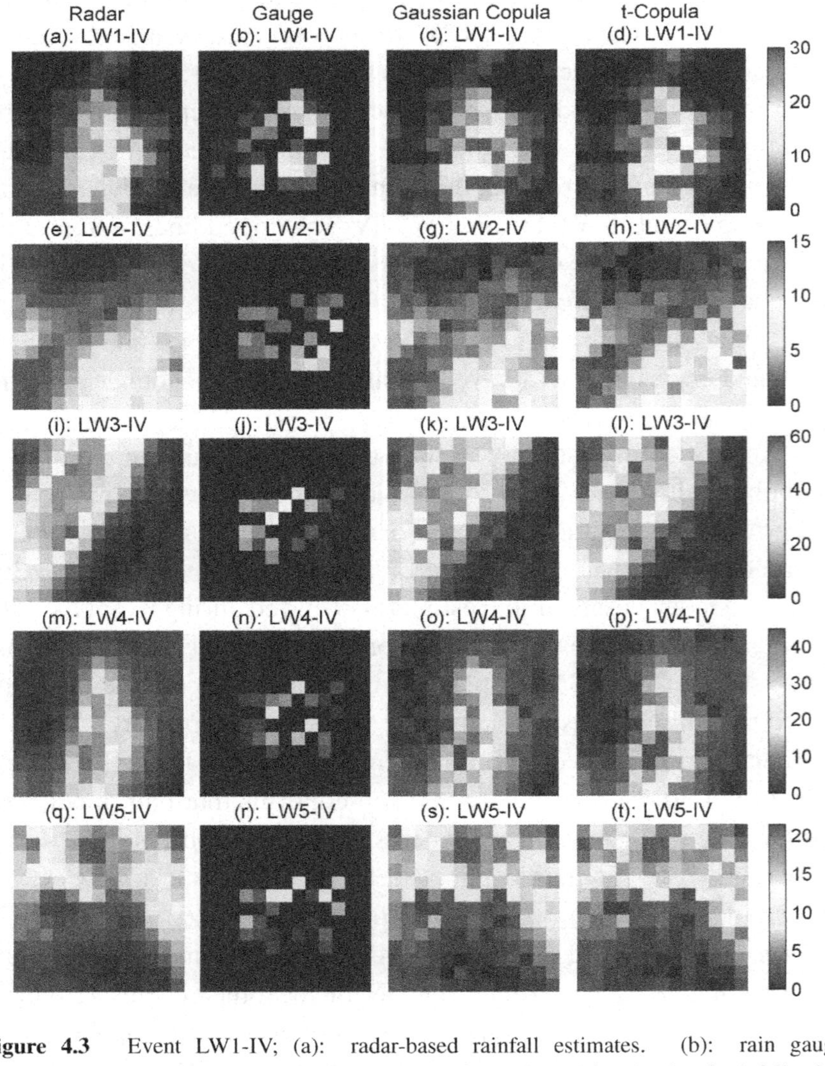

Figure 4.3 Event LW1-IV; (a): radar-based rainfall estimates. (b): rain gauge measurements. (c): simulated rainfall using Gaussian copula. (d): simulated rainfall using t-copula. Figures (e) to (h), (i) to (l), (m) to (p) and (q) to (t), display similar plots for Events LW2-IV, LW3-IV, LW4-IV and LW5-IV, respectively.

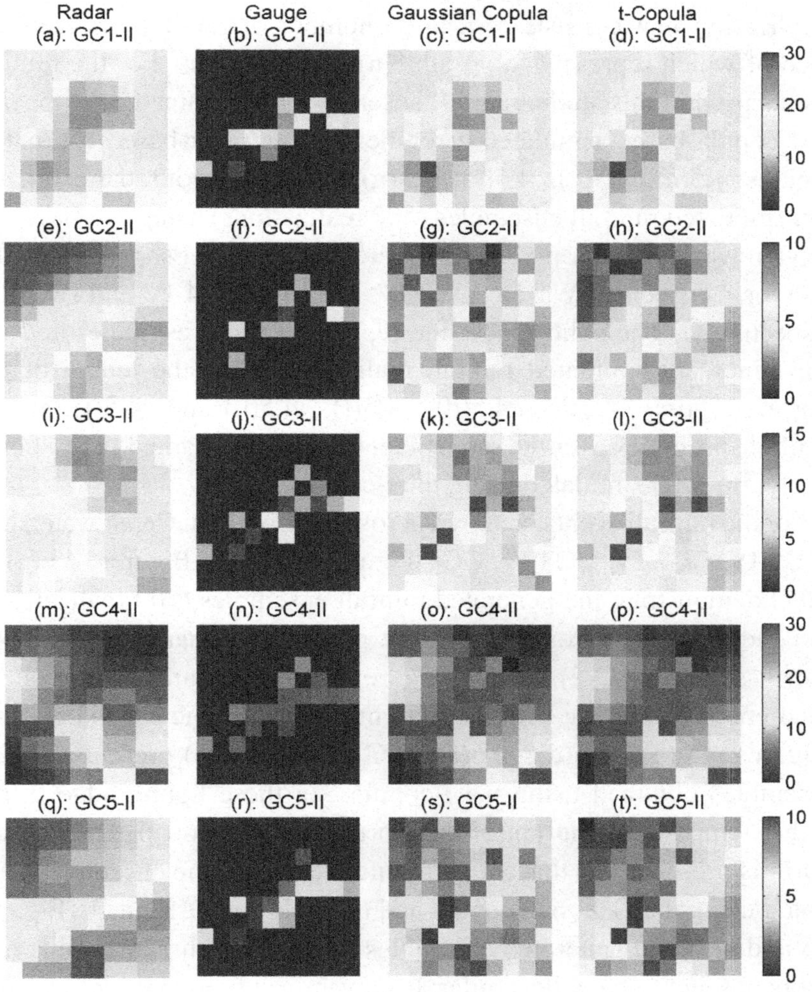

Figure 4.4 Event GC1-II; (a): radar-based rainfall estimates. (b): rain gauge measurements. (c): simulated rainfall using Gaussian copula. (d): simulated rainfall using t-copula. Figures (e) to (h), (i) to (l), (m) to (p) and (q) to (t), present similar plots for Events GC2-II, GC3-II, GC4-II and GC5-II, respectively.

As mentioned previously, in this study an ensemble approach is used to describe the uncertainty associated with rainfall estimates. A statistical ensemble of a random process (here, rainfall error) is an idealization that consists of a large number of random realizations, each of which represents a possible true observation. For the rainfall events used in this study, rainfall ensembles are obtained by imposing 500 copula-based simulated error fields over radar-based estimates. Figures 4.5(a), 4.5(c), 4.5(e), 4.5(g) and 4.5(i) plot radar estimates and simulated rainfall ensembles (500 realizations) using the Gaussian copula over one radar pixel (the square-marked pixel shown in Figure 3.1) for the Events LW1-IV, LW2-IV, LW3-IV, LW4-IV and LW5-IV, respectively. The solid black line represents radar estimates and the gray lines are simulated rainfall realizations over the length of the storm. Figures 4.5(b), 4.5(d), 4.5(f), 4.5(h) and 4.5(j) present rainfall ensembles over the same pixel (the square-marked pixel shown in Figure 3.1)) simulated using the t-copula.

For the rainfall events occurred across the Goodwin Creek watershed (GC1-IV, GC2-IV, GC3-IV, GC4-IV and GC5-IV), Figures 4.6(a) to 4.6(j) demonstrate the generated rainfall ensembles (500 realizations) and radar estimates over one radar pixel (the square-marked pixel shown in Figure 3.2). In this figure, the first column shows simulated ensembles using the Gaussian model, while the second column (Figures 4.6(b), 4.6(d), 4.6(f), 4.6(h) and 4.6(j)) presents rainfall ensembles obtained using the t-copula. In these figures, due to the higher temporal resolution of the storms, only a short period of each storm is presented so that the differences between the ensembles generated using the Gaussian copula and t-copula can be seen. In Figures 4.5 and 4.6, the generated ensembles using the Gaussian copula and t-copula seem to be quite similar; however, in subsequent sections, numerous statistical testes are used to highlight the differences between both models and also to compare them numerically.

In order to validate the presented models, multiple rainfall realizations were simulated with different numbers of gauges to investigate if the estimated uncertainty encompasses rain gauge measurements when less numbers of gauges were available. For the Event LW1-IV, Fig-

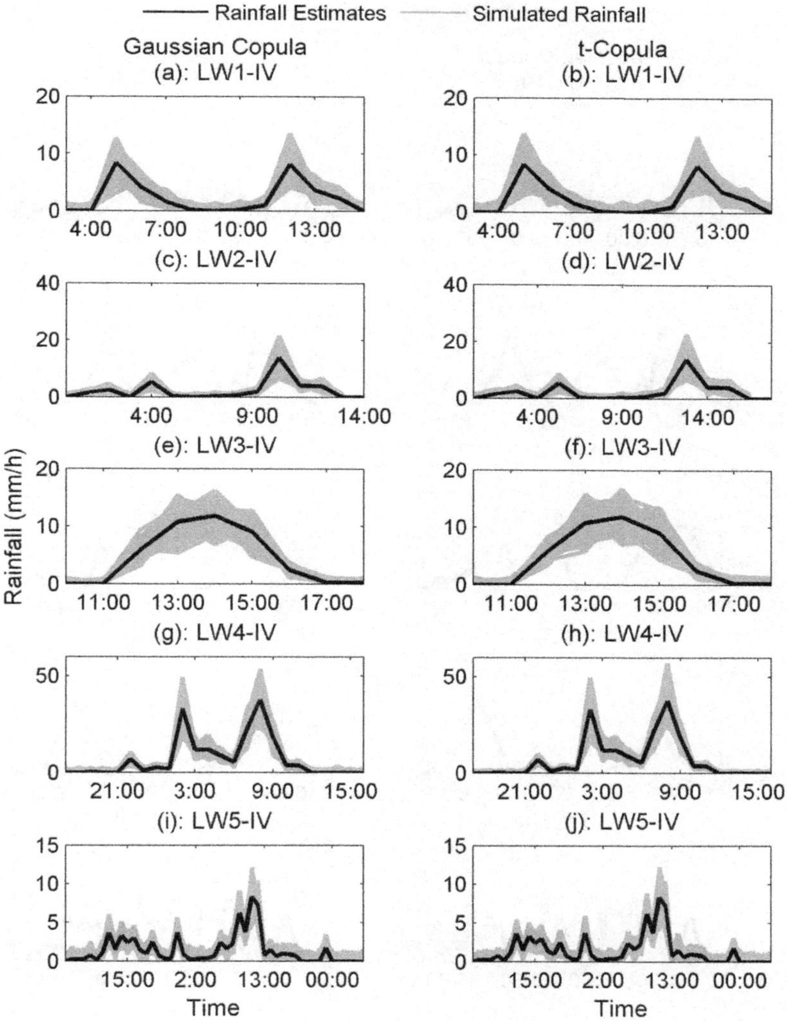

Figure 4.5 Events LW1-IV to LW5-IV: Rainfall ensembles (500 realizations) simulated using Gaussian copula ((a),(c),(e),(g),(i)) and t-copulas ((b),(d),(f),(h),(j)) over the square-marked pixel in Figure 3.1.

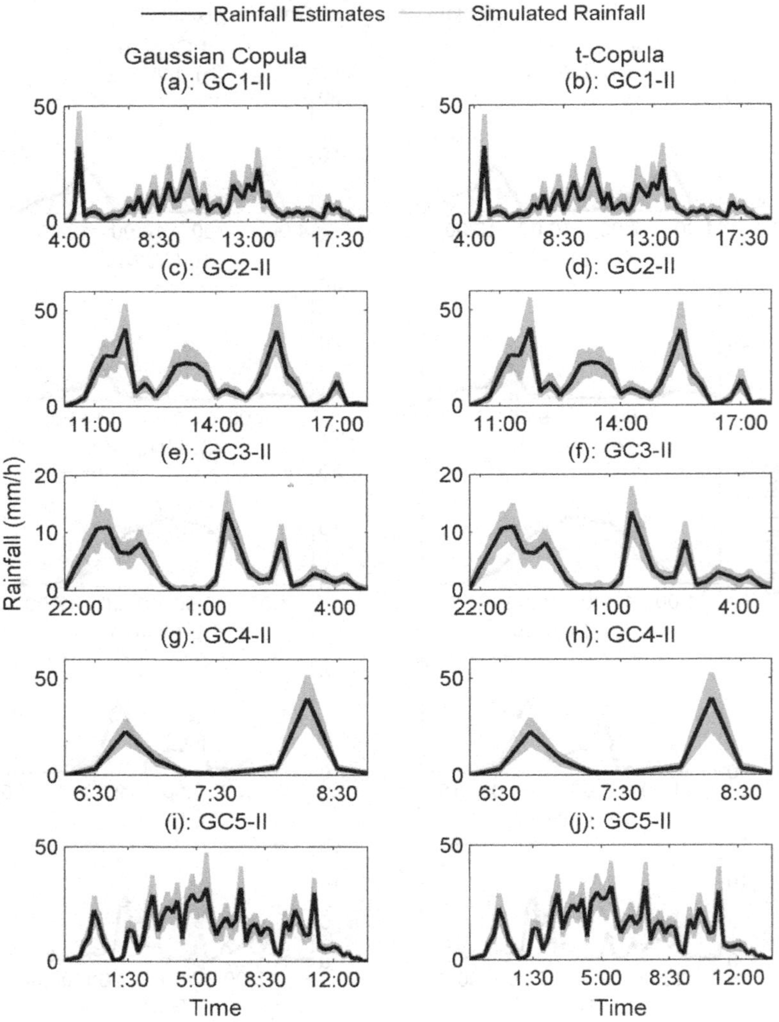

Figure 4.6 Events GC1-II to GC5-II: Rainfall ensembles (500 realizations) simulated using Gaussian copula ((a),(c),(e),(g),(i)) and t-copulas ((b),(d),(f),(h),(j)) over the square-marked pixel in Figure 3.2.

ures 4.7(a) to 4.7(c) display rainfall ensembles obtained using the Gaussian copula model over the x-marked pixel (shown in Figure 3.1) which contained one of the removed rain gauges. Figures 4.7(d) to 4.7(f) present simulated rainfall realizations obtained using the t-copula model over the same radar pixel. In Figures 4.7(a) and 4.7(d), all available gauges are included, whereas in Figures 4.7(b) and 4.7(e), 20 rain gauges (50% of the available data, representing approximately 1 gauge in 31 km^2) are used for simulations. The selected rain gauges are marked with triangles and diamonds in Figure 4.9(a). In Figures 4.7(c) and 4.7(f), only 5 gauges (12% of the available data representing approximately 1 gauge in 122 km^2), marked with diamonds in Figure 4.9(a), are used for ensemble generation. A similar figure is provided for Event GC3-II where rainfall ensembles are generated for the x-marked pixel, shown in Figure 3.2, using different number of gauges: 18 rain gauges shown in Figure 4.9(b); 10 rain gauges marked with triangles and diamonds in Figure 4.9(b); 5 rain gauges marked with diamonds in Figure 4.9(b). In these figures, the solid and dashed lines represent radar rainfall estimates and rain gauge measurements, whereas the gray lines are simulated radar rainfall realizations over the entire storm. With a few exceptions, the estimated uncertainties associated with the rainfall estimates reasonably enclose the ground reference measurements (see Figure 4.8).

In order to evaluate the rainfall ensembles, which are generated using different number of gauges, the number of time steps (n_{out}) where the rain gauge measurements fall outside the generated ensembles is counted and normalized with respect to the length of the storm. It is worth remembering that an ensemble of rainfall estimates is expected to encompass rainfall observations. Table 4.3 gives the number of time steps (n_{out} in percentage) that the estimated uncertainty did not enclose the rain gauge measurements. When all gauges are included, the estimated uncertainty for the Little Washita rainfall events (LW1-IV to LW5-IV) encompass the rain gauge measurements entirely. For the Goodwin Creek events (GC1-II to GC5-IV), the gauge measurements do not fall in the estimated ensemble in few time steps (most are less than 2 %). mainly resulting from: (a) the Little Washita data are multi-sensor radar estimates that are expected to be in better agreement with rain gauge measurements; (b) the temporal resolution of the Goodwin Creek rainfall events are higher. When the number of rain gauges is reduced to 50%, n_{out} is increased but not significantly (most remain less than 2.5 %). Further reduction of gauges to 12% for the Little Washita events and 25% for the Goodwin Creek events resulted in an increase of nearly 7% in some instances. The results indicate that even with a few rain gauges, the simulated ensembles obtained from both Gaussian and t-copula models, encompass approximately 95 % of the rain gauge measurements. Overall, the t-copula model performs slightly better (compare columns 4 and 6 with columns 5 and 7, respectively).

Table 4.3 The number of time steps, n_{out} (%), that the estimated uncertainty did not enclose the rain gauge measurements.

	All Gauges		50% LW Gauges 50% GC Gauges		12% LW Gauges 25% GC Gauges	
Event ID	Gaussian	t-copula	Gaussian	t-copula	Gaussian	t-copula
LW1-IV	0.0	0.0	2.1	2.3	4.4	3.9
LW2-IV	0.0	0.0	1.3	0.9	3.8	4.3
LW3-IV	0.0	0.0	1.4	1.2	4.1	3.0
LW4-IV	0.0	0.0	1.9	1.6	3.3	3.7
LW5-IV	0.0	0.0	2.3	1.8	4.2	4.0
GC1-II	0.5	0.3	2.1	2.0	4.9	4.4
GC2-II	1.6	1.0	2.5	2.3	6.1	6.0
GC3-II	1.3	1.2	2.4	2.1	5.7	5.5
GC4-II	1.4	1.1	2.8	2.4	5.6	5.2
GC5-II	1.6	1.3	2.5	2.5	6.7	6.4

It is pointed out that in the performed analyses with different sets of rain gauges, the model parameters were estimated based on the selected rain gauges using the method introduced in Section 4.3. In the following, the effects of reducing the number of gauges on the dependence structure and copula fitting are discussed.

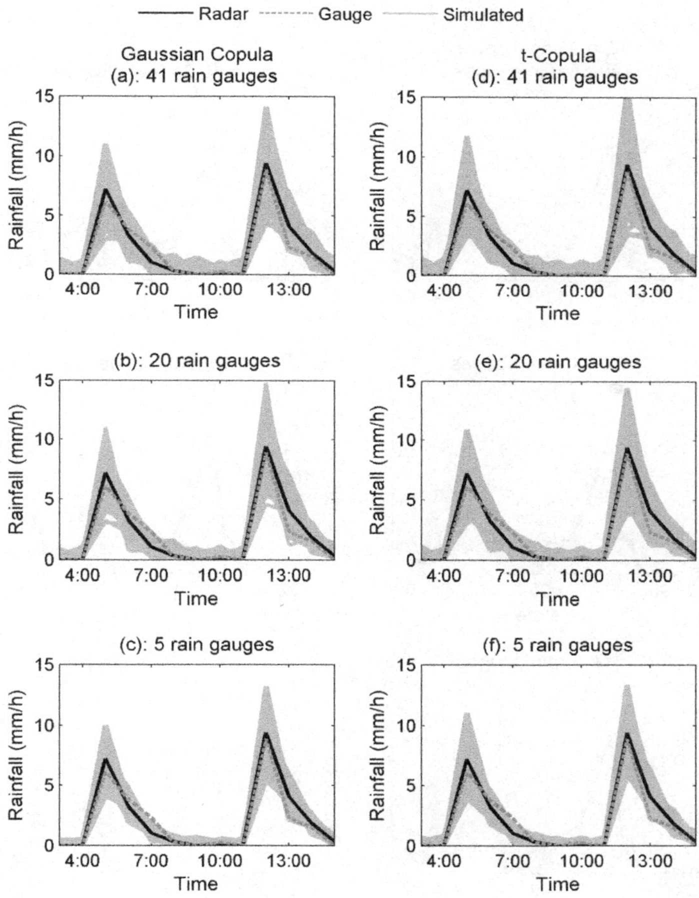

Figure 4.7 Event LW1-IV: Simulated rainfall ensembles using the Gaussian copula ((a: 41 gauges), (b: 20 gauges; marked with triangles and diamonds in Figure 4.9(a)) and (c: 5 gauges; marked with diamonds in Figure 4.9(a))) and t-copula ((d: 41 gauges), (e: 20 gauges) and (f: 5 gauges)).

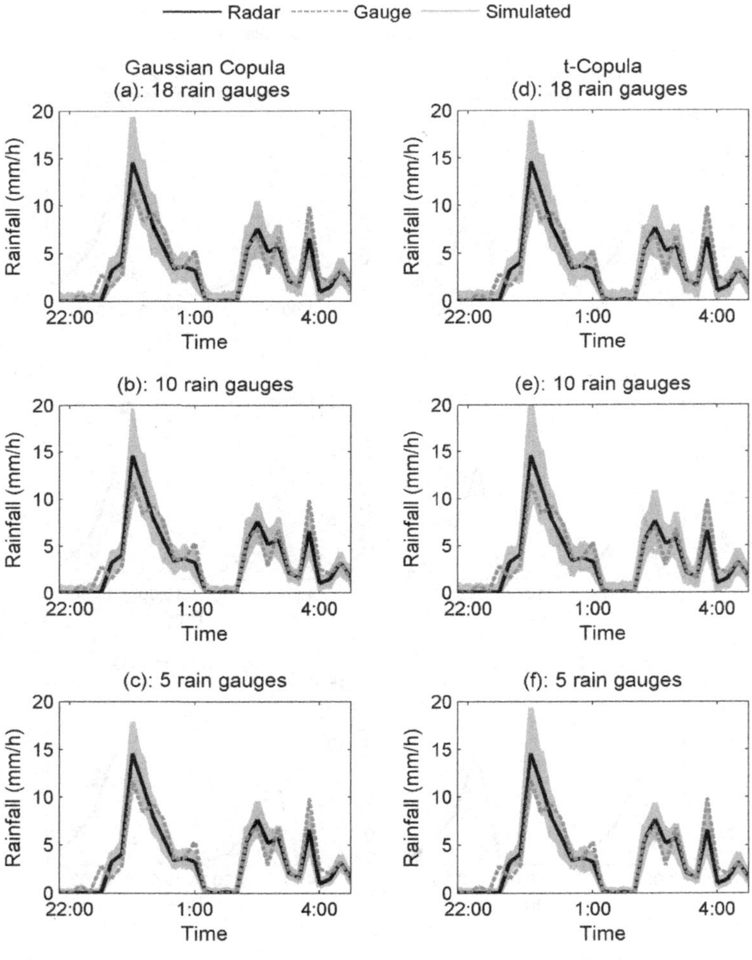

Figure 4.8 Event GC3-II: Simulated rainfall ensembles using the Gaussian copula ((a: 18 gauges), (b: 10 gauges; marked with triangles and diamonds in Figure 4.9(b)) and (c: 5 gauges; marked with diamonds in Figure 4.9)(b)) and t-copula ((d: 18 gauges), (e: 10 gauges) and (f: 5 gauges)).

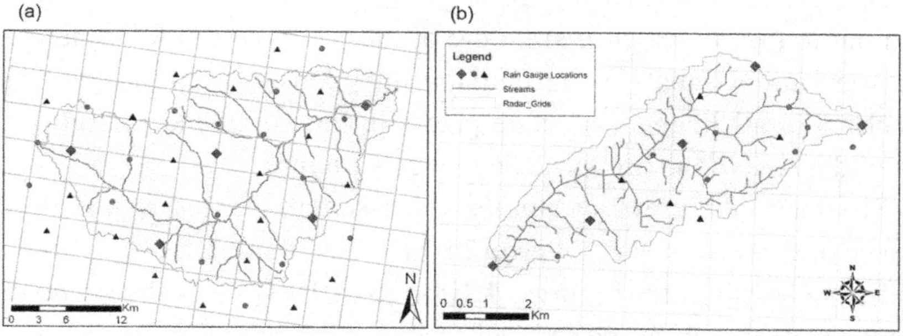

Figure 4.9 The rain gauges used for model verification: (a) Little Washita watershed; (b) Goodwin Creek watershed.

4.5 SPATIAL AND TEMPORAL DEPENDENCIES

The unavailability of a true representation of rainfall in space and time necessitates that most error models neglect the temporal and spatial dependencies inherent to radar rainfall error fields. For example, Smith et al. (1989) modeled random error assuming no temporal or spatial correlation beyond the size of one radar pixel (4 km x 4 km). However, significant spatio-temporal dependency may exist in radar error fields, particularly in space (Jordan et al. (2003)). Carpenter and Georgakakos (2004) recognized this issue and indicated the need to account for rainfall error dependencies. Jordan et al. (2003) introduced a random cascade model to generate error fields taking into account error correlations. The results showed non-negligible dependencies in the error field especially in space. Ciach et al. (2007) showed that the radar random error component was correlated in space and the estimated correlations were higher in the cold season. Habib et al. (2008) showed that the spatial dependence of radar error is not negligible and can have a significant effect on streamflow hydrologic predictions. Generally, one can argue that a reasonable representation of dependence structure may be required to generate an ensemble of rainfall data with the similar dependence structure as that of the rainfall estimates. However it may not be possible to obtain multiple generated rainfall fields with the exact correlation matrix as that of the original field. Nonetheless, the general dependence structure of simulated radar fields should be similar to the dependence structure of the observed rainfall fields. The reason for this is that if some neighboring pixels show high dependence in the rainfall image, the same pixels in the simulated rainfall fields should also exhibit similar dependence structure.

Similar to the previous studies discussed above, the spatial dependence of the observed radar error in the selected rainfall events is found to be non-negligible. Figures 4.10(a) to 4.10(f) show the correlation matrices of the observed for 6 of the selected rainfall events. One can see that the rank correlation values cannot be ignored in simulations. Therefore, the rainfall uncertainty (ϵ) is assumed to be correlated in space and the Gaussian copula and t-copula are used for describing the dependence structure. The Spearman rank correlation (ρ_s) matrix is employed to assess the dependence structure of the simulated error and rainfall fields. The Spearman rank correlation is a nonparametric method of describing the dependencies in terms of ranks and is independent of the marginal distributions (Spearman (1904); Hollander and Wolfe (1973)):

$$\rho_s = \frac{n\sum\chi_i\xi_i - \sum\chi_i\sum\xi_i}{\sqrt{n\sum\chi_i^2 - (\sum\chi_i)^2}\sqrt{n\sum\xi_i^2 - (\sum\xi_i)^2}} \tag{4.5}$$

where : χ_i = rank of x_i in X
ξ_i = rank of y_i in Y

For the Little Washita and Goodwin Creek rainfall events, Figures 4.11(left) and 4.12(left) present the Spearman correlation matrices of rainfall estimates for 10 radar pixels. Figures 4.11(middle) and 4.12(middle) show the correlation matrices of one set of simulated rainfall fields using the Gaussian model, whereas Figures 4.11(right) and 4.12(right) demonstrate the correlation matrices of one set of generated rainfall fields using the t-copula (after imposing generated rainfall error (ϵ) on rainfall estimates). As shown in the provided examples, the overall spatial dependence structure is reasonably preserved. In order

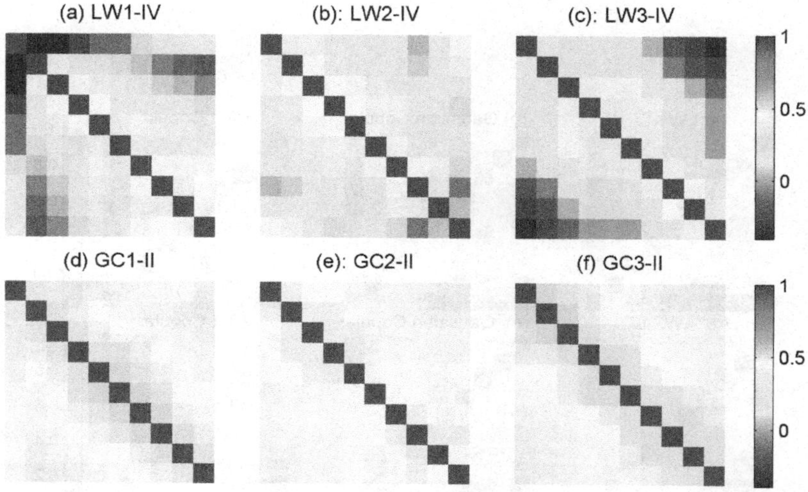

Figure 4.10 Rank correlation matrices of the observed radar error.

to analyze the correlation matrices of the simulated and observed radar fields and to quantify the error, the Mean Absolute Error (MAE) of the correlation matrices of the simulated radar estimates is calculated with respect to the observed radar estimates. The mean absolute error which is a quantity often used to measure how close simulations are to the observations, is defined as (Cox and Hinkley (1974); Bernardo and Smith (2000)):

$$MAE = \frac{1}{n}\sum_{i=1}^{n}|S_i - O_i| \tag{4.6}$$

where : S_i = simulated values
O_i = observed values
n = number of pairs (simulation, observation)

Table 4.4 lists the MAE values for the correlation matrices of the simulated rainfall fields with respect to the correlation matrix of the observed radar rainfall. In this table, the MAE is also provided for the cases where the number of gauges is reduced (see Section 4.4 for

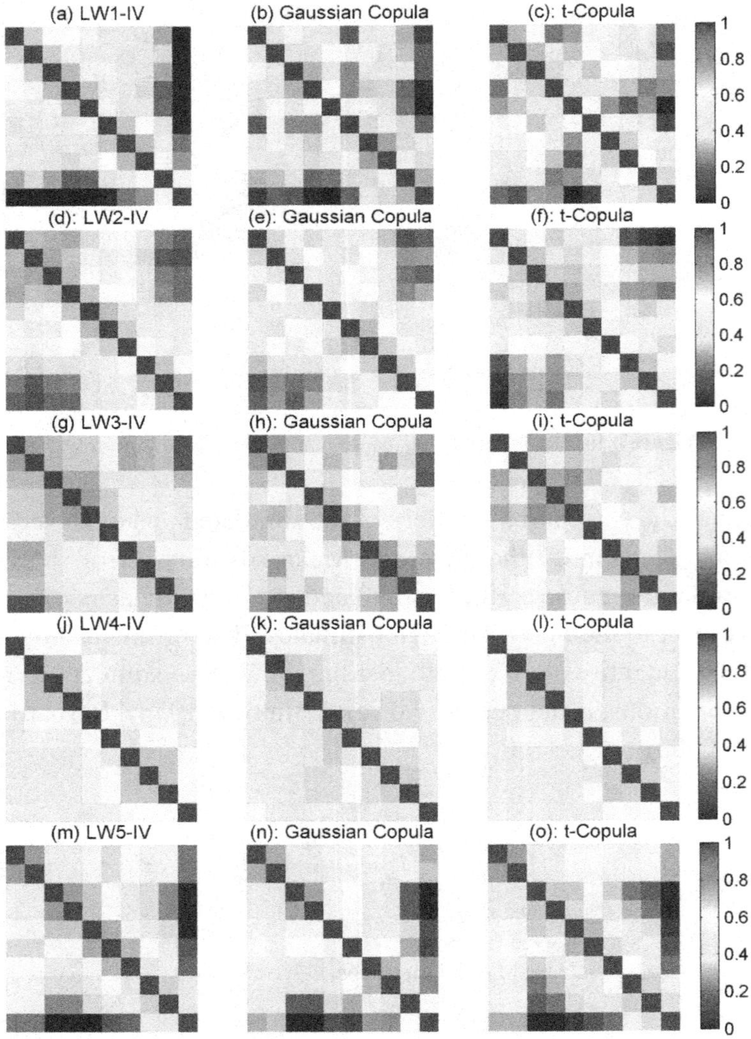

Figure 4.11 Rank correlation matrices of Events LW1-IV to LW5-IV (left); Rank correlation matrices of one set of simulated data using: Gaussian copula (middle); t-copula (right).

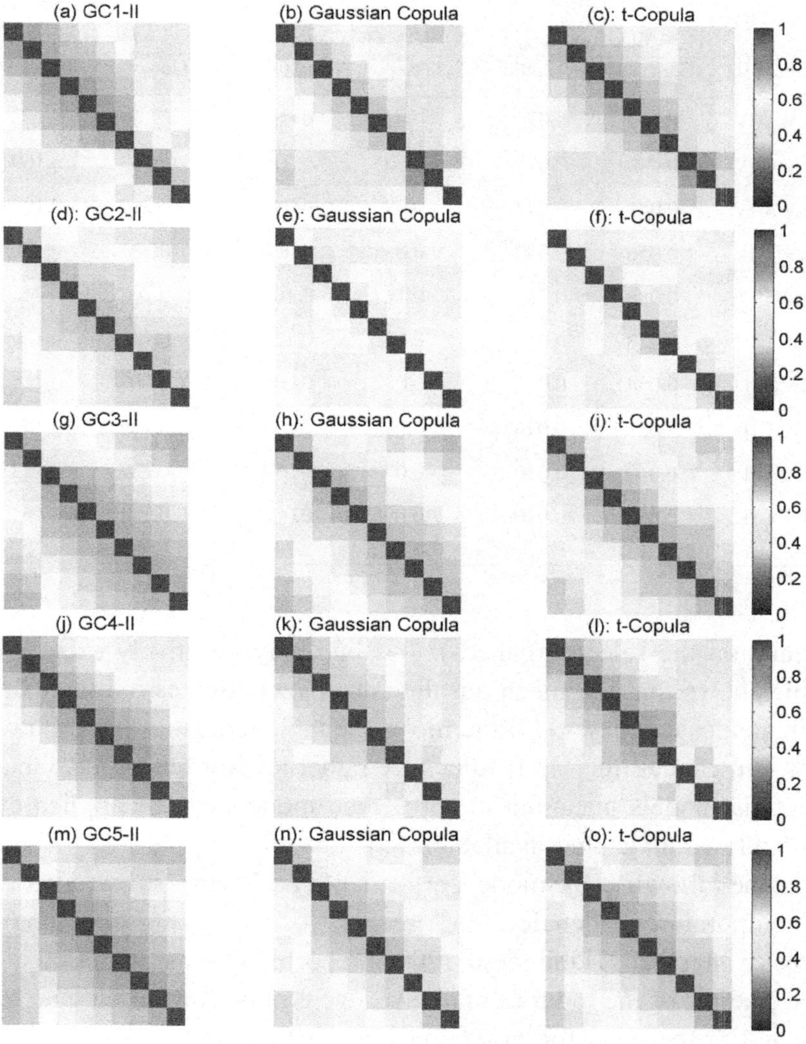

Figure 4.12 Rank correlation matrices of Events GC1-IV to GC5-IV (left); Rank correlation matrices of one set of simulated data using: Gaussian copula (middle); t-copula (right).

Table 4.4 The mean absolute error (MAE) of the correlation matrices of the simulated rainfall fields with respect to the observed.

	All Gauges		50% LW Gauges 50% GC Gauges		12% LW Gauges 25% GC Gauges	
Event ID	Gaussian	t-copula	Gaussian	t-copula	Gaussian	t-copula
LW1-IV	0.061	0.076	0.091	0.088	0.159	0.181
LW2-IV	0.040	0.039	0.105	0.093	0.183	0.169
LW3-IV	0.041	0.034	0.086	0.083	0.217	0.199
LW4-IV	0.056	0.030	0.055	0.061	0.133	0.130
LW5-IV	0.088	0.054	0.117	0.109	0.208	0.211
GC1-II	0.078	0.052	0.122	0.129	0.122	0.129
GC2-II	0.054	0.035	0.082	0.080	0.109	0.116
GC3-II	0.053	0.026	0.095	0.090	0.156	0.143
GC4-II	0.043	0.030	0.078	0.084	0.137	0.132
GC5-II	0.028	0.010	0.095	0.086	0.144	0.151

details on the selected gauges). As one may intuitively expect, the table shows that the mean absolute error increases as the number of rain gauges decreases. Furthermore, the table indicates that even with few rain gauges, the MAE values are rather low for both Gaussian and t-copula models, meaning the presented models can be applied even with a few rain gauges available.

In the following, the model performance is also investigated in terms of temporal dependencies. In a recent work involving the Goodwin Creek watershed, Habib et al. (2008) reported that the temporal autocorrelations of the radar rainfall error were rather low at the first time lag and close to zero for larger time lags. In the selected rainfall events, the temporal autocorrelations of the observed error were found to be negligible (e.g. see Figure 4.13) and hence the temporal autocorrelation of rainfall error is not explicitly included in the presented models. However, when simulated error fields are imposed over radar estimates, the resulting rainfall fields will be temporally correlated, which results from the dominance of the underlying autocorrelations of the rainfall estimates. Figures 4.14(left) and 4.15(left) demonstrate the temporal autocorrelations of the radar estimates and 500 simulated rainfall fields using the Gaussian copula for the square-marked pixel in Figures 3.1 and 3.2, respectively. Figures 4.14(right) and 4.15(right) plot the temporal autocorrelations when t-copula was used for simulations. The figures confirm that simulated rainfall realizations have realistic temporal autocorrelation characteristics. However, most of the simulated rainfall realizations have lower temporal autocorrelations, which result from the perturbation of the rainfall estimates with random fields.

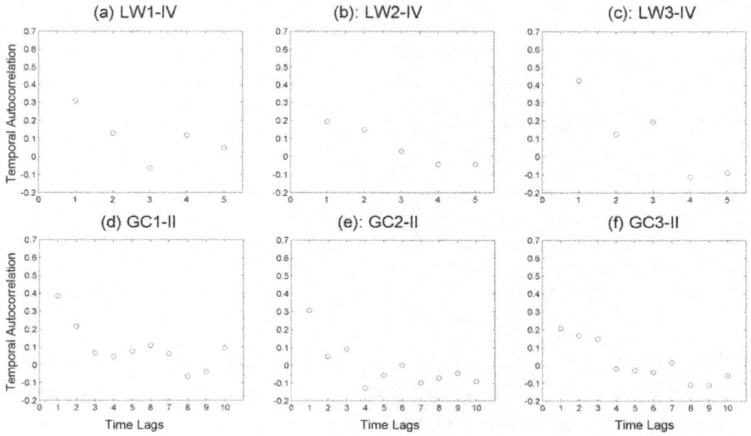

Figure 4.13 Temporal autocorrelation of the observed radar error for the x-marked pixel in Figures 3.1 and 3.2.

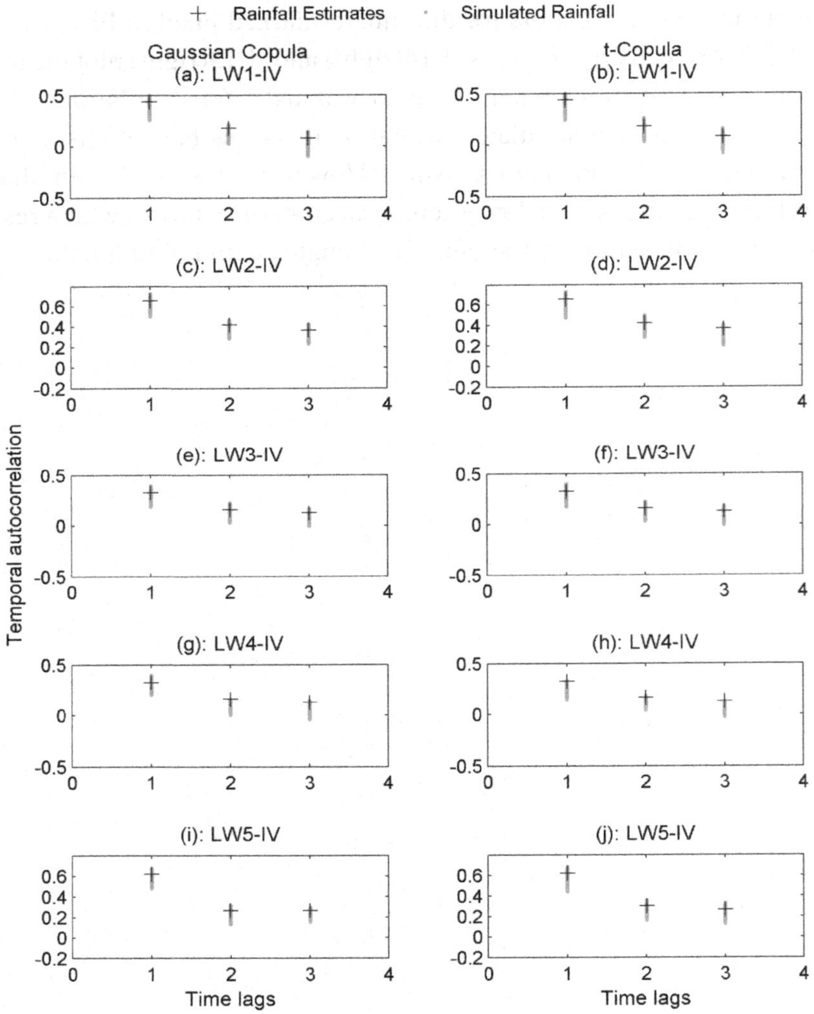

Figure 4.14 Temporal autocorrelation of rainfall estimates and 500 simulated rainfall realizations of Events LW1-IV to LW5-IV using: Gaussian copula (left); t-copula (right) for the square-marked pixel shown in Figure 3.1.

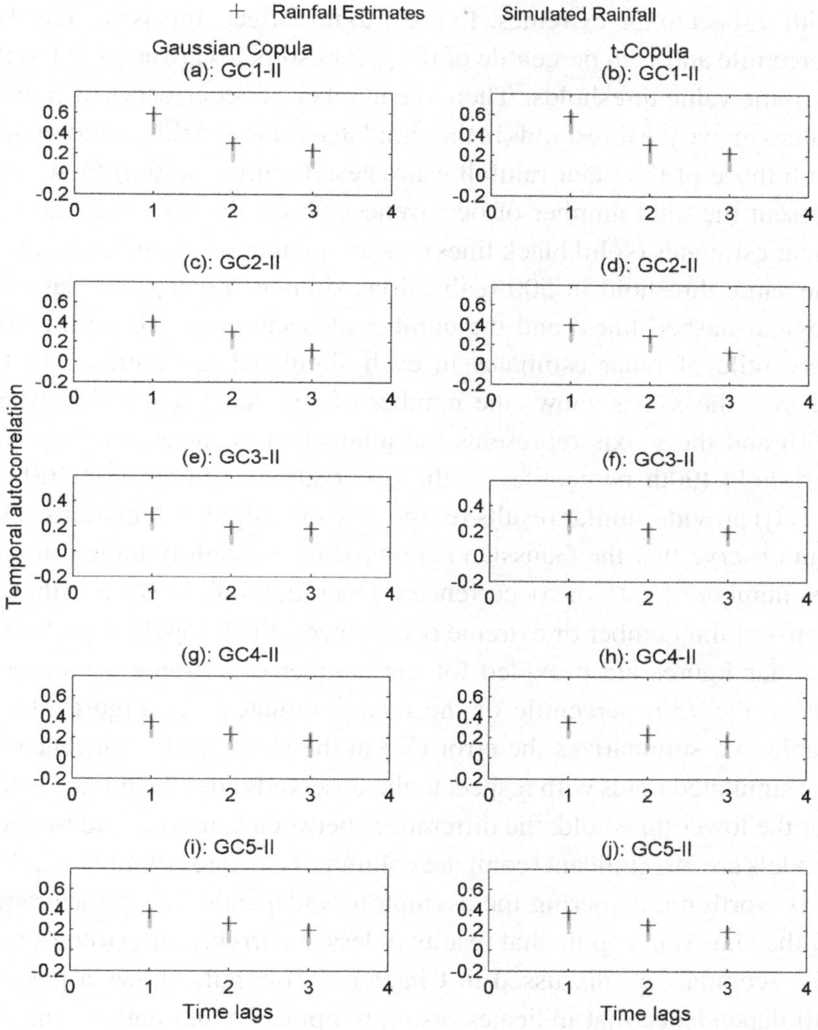

Figure 4.15 Temporal autocorrelation of rainfall estimates and 500 simulated rainfall realizations of Events GC1-IV to GC5-IV using: Gaussian copula (left); t-copula (right) for the square-marked pixel shown in Figure 3.2.

4.6 EXTREME VALUES

In this section, the Gaussian and t-copula based models are compared with respect to the extremes. In order to investigate this issue, the 90th percentile and 75th percentile of the radar estimates are assumed as the extreme value thresholds. Then, the number of occurrences of rainfall values above the thresholds in the simulated rainfall fields are compared with those of the radar rainfall estimates. Figures 4.16(a) to 4.16(e) present the total number of occurrences above the 90th percentile of radar estimates (solid black lines), mean number of occurrences above the same threshold in 500 realizations simulated using the Gaussian copula (dashed lines) and the number of occurrences above the 90th percentile of radar estimates in each simulated realization. In the figures, the x-axis shows the number of simulated realizations (here 500) and the y-axis represents the number of occurrences above the threshold (90th percentile) of the observations. Figures 4.16(f) to 4.16(j) provide similar results for the case of t-copula simulations. One can observe that the Gaussian simulated fields slightly underestimate the number of extreme occurrences. Overall, the figures show that, in terms of the number of extreme occurrences, the t-copula is preferred. Similar figures are provided for the number of extreme occurrences above the 75th percentile of the radar estimates (see Figure 4.17). Table 4.5 summarizes the error (%) in the number of occurrences in the simulated fields with respect to the observed rainfall data. As listed, for the lower threshold, the differences between Gaussian and t-copula models are insignificant (compare columns 4 and 5 to columns 2 and 3). It is worth remembering the asymptotic independence characteristics of the Gaussian copula that results in less occurrence of extremes than the t-copula. As discussed in Chapter 2, t-copula shows a positive tail dependence that indicates, using t-copula for simulation, one can expect larger extreme values to be generated simultaneously than by the Gaussian copula (Embrechts et al. (2001)). This implies that for modeling extremes, Gaussian copula may not be an appropriate choice of copula family.

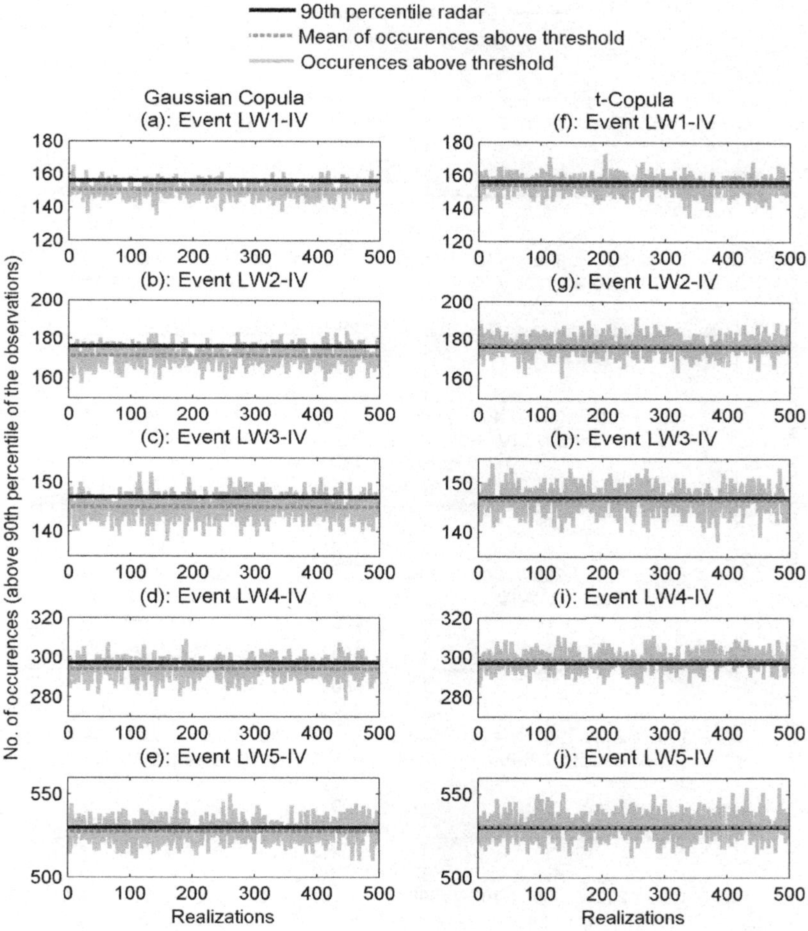

Figure 4.16 Total number of occurrences above the 90th percentile of radar data (solid black lines), mean number of occurrences above the same threshold in 500 realizations (dashed lines) and the number of occurrences above the 90th percentile of radar estimates in each simulated realizations.

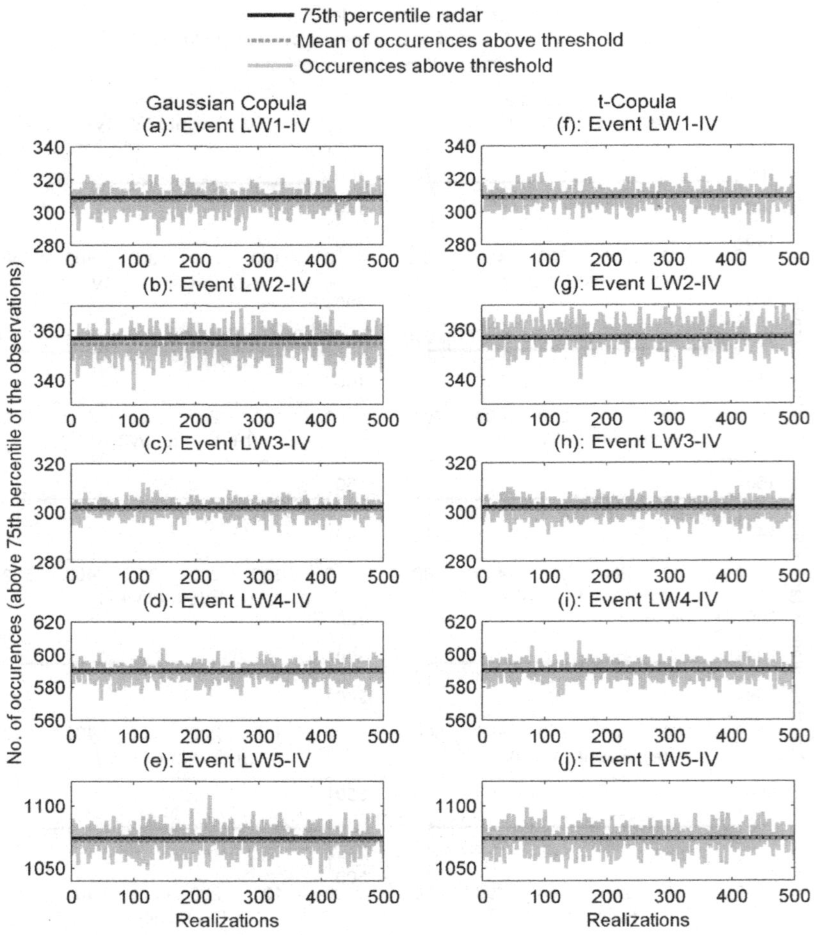

Figure 4.17 Total number of occurrences above the 75th percentile of radar data (solid black lines), mean number of occurrences above the same threshold in 500 realizations (dashed lines) and the number of occurrences above the 75th percentile of radar estimates in each simulated realizations.

Table 4.5 The error (%) in the number of extreme occurrences in the simulated fields with respect to the observed rainfall data.

	90th percentile		75th percentile	
Event ID	Gaussian	t-copula	Gaussian	t-copula
LW1-IV	3.6	1.9	0.7	0.6
LW2-IV	2.8	0.9	1.1	0.8
LW3-IV	1.4	0.6	0.3	0.3
LW4-IV	1.5	0.8	0.6	0.5
LW5-IV	0.7	0.6	0.9	0.5

4.7 SUMMARY AND DISCUSSION

Remotely sensed rainfall estimates, unlike traditional gauge measurements, capture the details of spatial variability of rainfall. However, they are subject to various errors from different sources that may affect hydrologic and meteorological applications. In this study two models were presented for simulation of rainfall error estimates using the well-known Gaussian copula and t-copula. The rainfall uncertainty is assumed to be an additive term as shown in Equation 4.1. A multiplicative error component was employed to account for the proportionality of rainfall error ($R_i \times \epsilon$) to rainfall rates reported in various studies (Habib et al. (2008); Ciach et al. (2007)). Multiple realizations of rainfall error are simulated based on the observed error values and proportional the magnitude of the rain rate for the corresponding pixel in rainfall estimates (radar image). This approach guarantees that simulated rainfall error fields match observed error data conditioned on rainfall rates. That is, one can avoid unrealistically large errors while the magnitude of rain rate is not significant. To demonstrate the model performance, Stage IV radar rainfall and Level II reflectivity data were used as the input to the models to generate an ensemble of rainfall realizations.

The copula parameters were estimated based on the available observations. In order to test how well the copula-based models fit the observations, cross validation was employed considering the mean absolute error (MAE) as the estimator. Cross-validation is a well known approach that can be used to evaluate fitted models to observations and also to compare performances of different predictive modeling procedures (Picard and Cook (1997); Efron and Tibshirani (1990)). Table 4.6 lists the results of the repeated random sub-sampling cross validation (Picard and Cook (1997)) for the fitted Gaussian copula and t-copula to the observations. It is noted than for the case where only 5 rain gauges were included in simulations, a different cross validation technique, leave-one-out cross-validation (LOOCV) was used due to the limited observations (see columns 6 and 7 in Table 4.6). In this approach a single observation from the original sample is used for validation, and the remaining observations as the training data. This is repeated for the number of observations such that each observation is used once as the validation data. One can see that the t-copula model fits better to the observations. Furthermore, the table indicates that with less number of gauges available, the mean absolute error does not change dramatically.

In the presented models, instead of fitting a standard distribution function to the data, the empirical CDF of observed error is applied to the uniformly simulated error values so that simulated error values will have the same CDF as that of observed. This significant improvement helps avoiding the problem of overshooting (simulating unrealistically large values) which typically happens when a standard distribution function is fitted and used for simulations. Visual comparisons and also the Two-sample Kolmogorov-Smirnov test confirmed that the CDFs of observed error were reasonably reproduced regardless of the type of copula family (here, Gaussian copula and t-copula) used for simulations.

In terms of preserving spatial dependencies, both Gaussian and t-copula models were quite similar. However, for simulation of extremes (rainfall error above/below a certain threshold), using t-copula seemed to have significant advantages over Gaussian copula. The results showed that for a constant threshold, occurrence of extremes in

Table 4.6 The results of cross validation for the Gaussian copula and t-copula models (MAE estimator).

	All Gauges		50% LW Gauges 50% GC Gauges		12% LW Gauges 25% GC Gauges	
Event ID	Gaussian	t-copula	Gaussian	t-copula	Gaussian	t-copula
LW1-IV	0.98	0.97	1.34	0.93	2.49	2.35
LW2-IV	0.85	0.84	1.31	1.23	2.50	2.28
LW3-IV	1.34	1.18	1.59	1.46	2.70	2.56
LW4-IV	1.03	0.93	1.07	1.00	2.47	2.20
LW5-IV	1.17	1.16	1.74	1.62	2.74	2.43
GC1-II	1.00	0.99	1.14	1.16	2.78	2.48
GC2-II	0.84	0.87	1.79	1.77	2.93	2.59
GC3-II	1.08	0.87	1.66	1.65	2.80	2.71
GC4-II	0.85	0.94	1.80	1.67	3.06	2.52
GC5-II	1.21	1.03	2.06	1.97	2.67	2.64

Gaussian copula is less than the t-copula. This implies that for modeling extreme values, the t-copula is more appropriate, as the Gaussian copula may underestimate occurrence of extremes. However, one should be aware that the choice of copula family may be problem dependent and cannot be simply generalized with few case studies.

The case studies and the results presented here indicate that one can simulate an ensemble of rainfall realizations by imposing generated error fields over rainfall estimates. The idea of perturbing rainfall estimates with error fields to obtain a rainfall ensemble, has some advantages over simulation of rainfall fields directly. For example, in the introduced models, accounting for temporal self-correlations of rainfall error was not included in the model explicitly. However, simulated rainfall fields, after perturbation with generated error fields, were found to be temporally autocorrelated, resulting from the dominance of the underlying temporal self-correlations of rainfall estimates.

In the following chapters, after introducing two different models for radar rainfall simulation, a comprehensive discussion is provided to address various issues including, the choice of copula, statistical characteristics of radar error and modeling spatial dependence structure.

CHAPTER 5

SIMULATION USING V-COPULA

5.1 INTRODUCTION

In the previous chapter, two elliptical copulas (the Gaussian copula and t-copula), were used for simulation of radar rainfall error fields. Elliptical copulas describe the dependence structure fully symmetrically (Abdous et al. (2003)). That is, they are restricted to radial asymmetry in the dependence structure. Previous studies, however, showed that the dependence structure may depart from Gaussianity (Bárdossy (2006); Journel and Alabert (1989); Gomez-Hernandez and Wen (1998)). Alternative approaches are required to account for asymmetry in the dependence structure. To address this issue, a multivariate non-Gaussian copula-based model is introduced that can account for asymmetrical dependencies through a number of parameters. In this model, the v-transformed copula (Bárdossy (2006)) is used to describe the dependencies. The v-transformed copula (v-copula) is expected to model data with pairwise different degrees of association. In other words, the desirable property of v-copula is the ability to highlight the difference between the dependence of high and low values. The general formulation of the v-transformed copula is discussed in Chapter 2. In this chapter, implication of this copula family for simulation of remotely sensed rainfall fields is discussed. As mentioned previously, remotely sensed rainfall estimates are subject to different types of error such as measurement biases, inherent retrieval errors and sampling uncertainty. (Hossain and Huffman (2008); Margulis et al. (2006); Dong et al. (2005); Lucieer and Kraak (2004); Bastin et al. (2002)). An ensemble-based approach is presented here for simulation of rainfall fields using the v-transformed copula. Using this model, multiple radar error fields are stochastically simulated and then imposed over rainfall estimates in order to generate and ensemble of rainfall realizations. The presented model is implemented for both radar reflectivity (Level II) and Stage IV radar-based multi-sensor precipitation estimates (the same events used in the previous chapter). In the following sections, after a brief discussion on the definition of symmetric and asymmetric dependence structure, the model is described and implemented for the selected rainfall events (introduced in Chapter 3). Several statistical tests are then performed to evaluate the applicability of the model.

5.2 MODEL DESCRIPTION

As mentioned before, one way to account for uncertainties associated with rainfall estimates is to generate an ensemble of rainfall realizations that is a representation of possible variabilities in precipitation data. Multiple realizations of rainfall fields can be obtained by imposing random error fields over the radar rainfall estimates: $R_s = R_i + R_i \times \epsilon$. The second term represents the error component which is known to be proportional to the magnitude of rain rate (Habib et al. (2008); Villarini et al. (2009); Ciach et al. (2007)). In this model, a multiplicative error component $R_i \times \epsilon$, as explained in Section 4.2, is included to account for the proportionality of simulated error estimates to the magnitude of rainfall rates. This property of radar-based rainfall error estimates is discussed in Chapter 4 in more detail.

In the presented model, the empirical cumulative distribution function (CDF) of the observed error is applied to the simulated uniform values so that the simulated fields are similar to the observations with respect to the CDF (for details the reader is referred to Chapter 4). It should be noted that in this model simulated error fields are conditioned on observed error values, which are typically obtained from rain gauge observations. By conditioning simulated error fields on observed error values and imposing error fields on radar estimates, the ground reference measurements will be honored at their locations in the simulated rainfall fields.

5.3 PARAMETER ESTIMATION

Using the available rain gauge measurements and radar estimates, the parameters of the v-copula model are to be estimated first. The v-transformed copula is parameterized with the correlation matrix and two other parameters, k and m, which are used to transform the multivariate normal copula (see Section 2.2.3 for details). In this study the parameter estimation method is based on Bárdossy and Li (2008) whereby the correlation matrix is described as a parameter dependent on different distance vectors (d). Based on this assumption, the correlation matrix can be obtained using the spherical function. A likelihood approach is used where the likelihood function itself consists of the product of multiple likelihood functions obtained from disjoint subsets of data. For each subset of observations, the likelihood function (L) for n observation points and m subsets can be expressed as:

$$L((m,k,s_0,s_1,s_2,r_1,r_2)|F(x_1),\ldots,F(x_n)) = \Pi_{i=1}^{m} c(S_i,(m,k,s_0,s_1,s_2,r_1,r_2)) \quad (5.1)$$

where : k, m = copula parameters
s_1, s_2 = sills of the spherical function
s_0 = nugget effect
r_1, r_2 = ranges of the spherical function
S_i = a subset of observations

Table 5.1 lists the parameters (s_0, s_1, s_2, r_1, r_2, k and m) of the v-transformed copula used for simulations. For Events LW3-IV and GC1-II, the parameter k was estimated to be 1, indicating a symmetric transformation function. It is noted that for the v-copula parameter estimation, similar to the approach explained in Chapter 4, estimation is performed based on the seasonal data (e.g. summer 2003, summer 2004, etc.) in which the event occurs.

Table 5.1 The parameters of the spherical function.

Event ID	s_0	s_1	s_2	r_1 [m]	r_2 [m]	k	m
LW1-IV	0.01	0.14	0.43	3400	14000	1.2	2.0
LW2-IV	0.03	0.12	0.41	2800	14500	0.9	3.1
LW3-IV	0.01	0.11	0.53	3800	14000	1.0	1.5
LW4-IV	0.04	0.19	0.45	3900	15500	2.1	3.8
LW5-IV	0.05	0.20	0.42	3400	13500	1.6	1.4
GC1-II	0.04	0.31	0.31	3200	7000	1.0	2.2
GC2-II	0.02	0.21	0.46	2600	5500	1.1	2.6
GC3-II	0.02	0.21	0.46	2600	5500	1.1	2.6
GC4-II	0.02	0.21	0.46	2600	5500	1.1	2.6
GC5-II	0.01	0.14	0.34	2400	6500	1.4	1.9

5.4 ENSEMBLE GENERATION USING V-COPULA

Having estimated the parameters, multiple random fields are simulated using the v-transformed copula as described in Chapter 2. The marginal of the observed error is then applied to the simulated univariate random fields using the Sklar Theorem (Equation 2.3). Figure 5.1(a) shows a radar rainfall image from rainfall Event LW1-IV, and Figure 5.1(b) presents the corresponding rain gauge measurements. Figure 5.1(c) displays an example of conditionally simulated rainfall field using the v-copula after imposing error fields over radar estimates. For rainfall Events LW2-IV to LW5-IV, Figures 5.1(d) to 5.1(o) show observed and simulated radar fields using the v-copula model. One can see that the model reproduces the rain gauge measurements (Figures 5.1(b), 5.1(e), 5.1(h), 5.1(k) and 5.1(n)) at their locations (see Figures 5.1(c), 5.1(f), 5.1(i), 5.1(l), 5.1(o)). During Events LW1-IV and LW2-IV, 42 and 22 rain gauges were available for simulations, respectively. While during the other events (LW3-IV, LW4-IV and LW5-IV), 20 rain gauges were in operation and available for analysis. Figures 5.2(a) to 5.2(o) display examples of simulated fields for the rainfall events that occurred across the Goodwin Creek watershed. In this figure, similar to Figure 5.1, the third column shows a number of simulated fields using the v-copula model (see Figures 5.2(c), 5.2(f), 5.2(i), 5.2(l) and 5.2(o)).

In this study the estimated uncertainty is described through an ensemble of rainfall estimates. The following displays the results of using the v-copula model to simulate rainfall ensembles for the selected events. Figures 5.3(a), 5.3(c), 5.3(e), 5.3(g) and 5.3(i) display radar estimates and simulated rainfall ensembles (500 realizations) over one radar pixel (the square-marked pixel shown in Figure 3.1) for rainfall Events LW1-IV, LW2-IV, LW3-IV, LW4-IV and LW5-IV, respectively. Figures 5.3(b), 5.3(d), 5.3(f), 5.3(h) and 5.3(j), on the other hand, plot simulated rainfall ensembles over the square-marked pixel shown in Figure 3.2 for rainfall Events GC1-II, GC2-II, GC3-II, GC4-II and GC5-II, respectively. In the figure, the solid black lines show radar estimates, whereas the gray lines represent simulated rainfall realizations over the length of the storm. As shown, the simulated ensembles do not show unrealistic realizations. In subsequent sections, various statistical tests are performed to verify the model performance in more detail. It is noted that in the v-copula model, the multiplicative error component

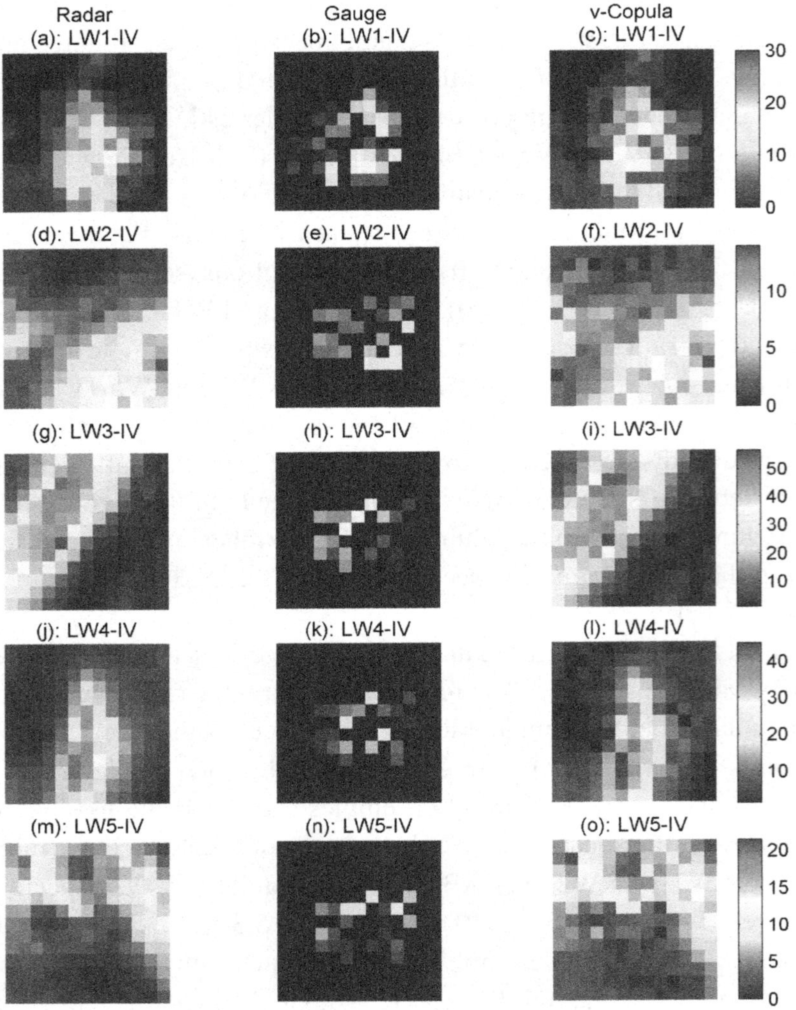

Figure 5.1 Event LW1-IV; (a): radar-based rainfall estimates. (b): rain gauge measurements. (c): simulated rainfall using the v-copula. Figures (d) to (f), (g) to (i), (j) to (l) and (m) to (o), display similar plots for Events LW2-IV, LW3-IV, LW4-IV and LW5-IV, respectively.

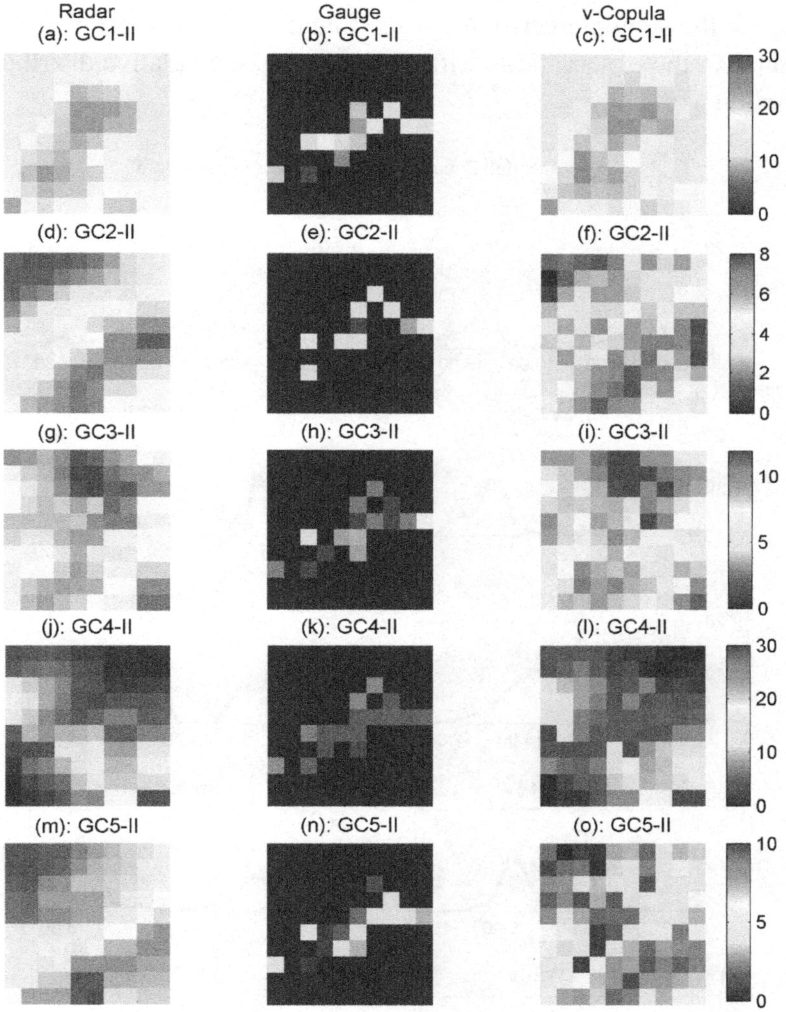

Figure 5.2 Event GC1-II; (a): radar-based rainfall estimates. (b): rain gauge measurements. (c): simulated rainfall using the v-copula. Figures (d) to (f), (g) to (i), (j) to (l) and (m) to (o), display similar plots for Events GC2-II, GC3-II, GC4-II and GC5-II, respectively.

(introduced in Chapter 4) is used to assure that the simulated error values are proportional to the rain rate magnitude. Furthermore, the empirical cumulative distribution function of the observed error is applied to the simulated error to ensure that the simulated error fields are similar to the observations with respect to the cumulative distribution function.

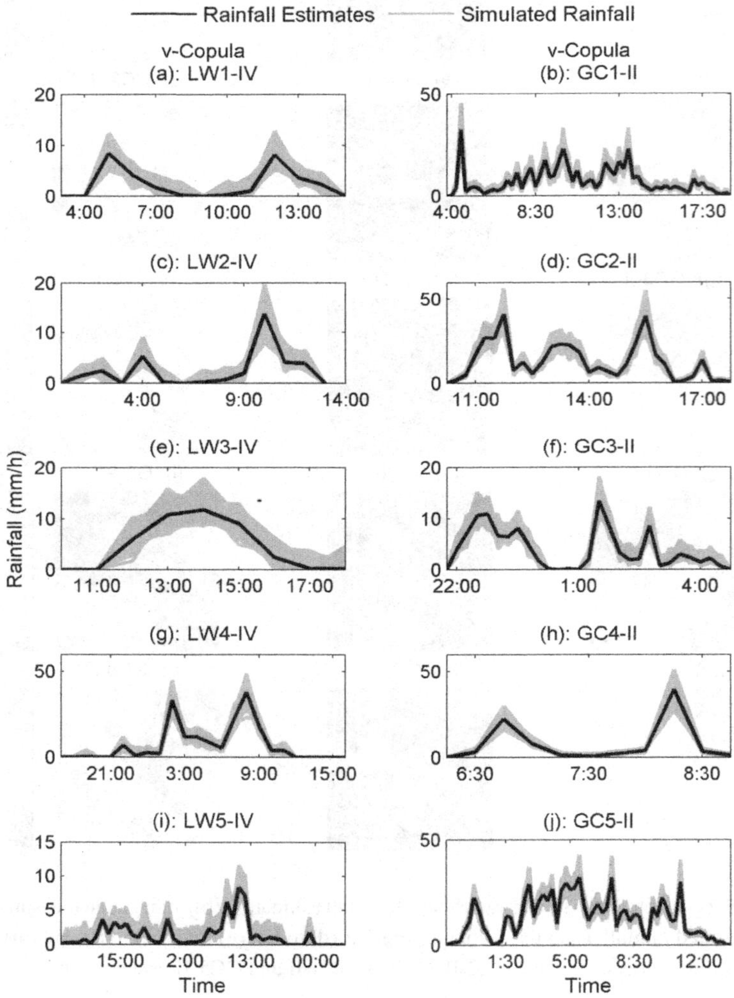

Figure 5.3 Simulated rainfall ensembles (500 realizations) using the v-copula for: the Little Washita events ((a),(c),(e),(g),(i)); the Goodwin Creek events ((b),(d),(f),(h),(j)).

The presented v-copula model is validated by removing 50% (22 rain gauges) and 88% (37 rain gauges) of the available data across the Little Washita watershed such that each gauge represents 31 km^2 and 122 km^2, respectively. For the Goodwin Creek rainfall events (GC1-II, GC2-II, GC3-II, GC4-II and GC5-II), 50% (10 rain gauges) and 75% (15 rain gauges) are removed from the analysis and the rainfall ensembles are simulated based on the remaining data. The idea is to investigate whether or not the estimated uncertainty encompasses rain gauge measurements if fewer gauges were available. For the Event LW1-IV, for example, Figures 5.4(a), 5.4(b) and 5.4(c) display rainfall ensembles simulated using the v-copula based model for the x-marked pixel (shown in Figure 3.1), which contained one of the removed rain gauges. In Figure 5.4(a) 41 rain gauges were used for simulation, whereas in Figures 5.4(b) and 5.4(c), 20 and 5 rain gauges were included for ensemble generation, respectively. Figures 5.4(d), 5.4(e) and 5.4(f) plot rainfall ensembles for Event GC3-II over the x-marked pixel, shown in Figure 3.2, using a different number of gauges; 5.4(d): 18 rain gauges shown in Figure 4.9(b); 5.4(e): 10 rain gauges marked with triangles and diamonds in Figure 4.9(b); 5.4(f): 5 rain gauges marked with diamonds in Figure 4.9(b). The solid and dashed lines represent radar estimates and rain gauge measurements, whereas the gray lines are simulated radar realizations (uncertainty) over the length of the storm. As shown, the estimated radar rainfall uncertainties reasonably encompass the rain gauge measurements with the exception of a few time steps. In general, a reasonable estimate of uncertainty (here, the generated ensemble) is expected to enclose the ground reference measurements.

The generated ensembles using different numbers of gauges are evaluated by counting the number of time steps (n_{out}) where the ground reference measurements do not fall inside the estimated uncertainty. Table 5.2 lists the number of time steps (n_{out}) in percentage. For the Little Washita events, the estimated uncertainty based on all the available gauges encloses the rain gauge estimates thoroughly. However, in Events GC1-II to GC5-IV, the estimated uncertainties do not encompass the ground reference measurements in few time steps (see the second column in Table 5.2). As explained in the previous chapter, the multi-sensor radar estimates (Little Washita Events) are expected to be in better agreement with rain gauge measurements. Reducing the available gauges to half results in an increase in the n_{out} of approximately 1% to 2% for both Goodwin Creek and Little Washita events. Further reduction of gauges to 25% for the Goodwin Creek events and 12% for the Little Washita events, results in an increase of n_{out} to a maximum of 6.5%. Overall the table indicates that with a few rain gauges across the watershed, the estimated uncertainty obtained from the v-copula model reasonably encloses the ground reference measurements. In Chapter 8, the estimated uncertainties obtained from the v-copula, Gaussian and t-copula models are compared numerically and are followed by a detailed discussion.

Table 5.2 The number of time steps, n_{out} (%), that the estimated uncertainty using the v-copula model did not enclose the rain gauge measurements.

	All Gauges	50% LW Gauges 50% GC Gauges	12% LW Gauges 25% GC Gauges
LW1-IV	0.0	2.2	4.2
LW2-IV	0.0	1.1	4.0
LW3-IV	0.0	1.2	4.1
LW4-IV	0.0	1.7	3.3
LW5-IV	0.0	2.2	4.3
GC1-II	0.4	2.3	4.6
GC2-II	1.4	2.5	6.1
GC3-II	1.2	2.3	5.5
GC4-II	1.2	2.6	5.5
GC5-II	1.5	2.6	6.5

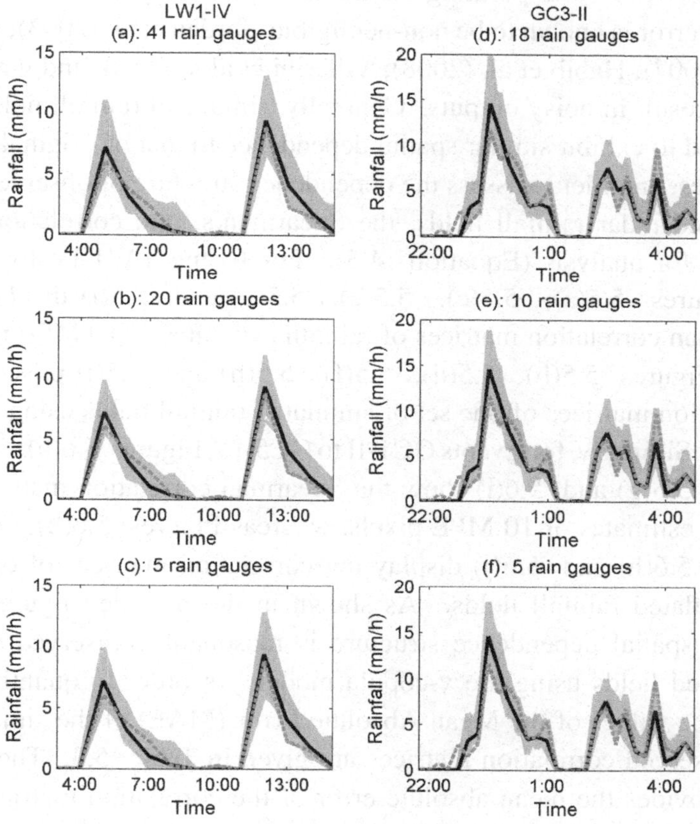

Figure 5.4 Simulated rainfall ensembles using the v-copula model: ((a: 41 gauges), (b: 20 gauges; marked with triangles and diamonds in Figure 4.9(a)) and (c: 5 gauges; marked with diamonds in Figure 4.9(a))); ((d: 18 gauges), (e: 10 gauges; marked with triangles and diamonds in Figure 4.9(b)) and (f: 5 gauges; marked with diamonds in Figure 4.9(b)).

5.5 SPATIAL AND TEMPORAL DEPENDENCIES

In this study, the error term (ϵ in Equation 4.1) is assumed to be correlated in space, and the spatial dependence is described using the v-copula. It is worth pointing out that the spatial dependence of radar rainfall error is known to be non-negligible (Jordan et al. (2003); Ciach et al. (2007); Habib et al. (2008); Villarini et al. (2009)), and ignoring it may result in noisy outputs. Generally, simulated rainfall fields are expected to exhibit similar spatial dependence to that of original radar estimates. In order to assess the dependence structure of observed and simulated radar rainfall fields, the Spearman's rank correlation (ρ_s) is used for analysis (Equation 4.5). For Events LW1-IV to LW5-IV, Figures 5.5(a), 5.5(c), 5.5(e), 5.5(g) and 5.5(i) display the Spearman correlation matrices of rainfall estimates of 10 MPE pixels, while Figures 5.5(b), 5.5(d), 5.5(f), 5.5(h) and 5.5(j) present the correlation matrices of one set of simulated rainfall fields using the v-copula. Similarly, for Events GC1-II to GC5-IV, Figures 5.6(a), 5.6(c), 5.6(e), 5.6(g) and 5.6(i) show the Spearman correlation matrices of rainfall estimates of 10 MPE pixels, whereas Figures 5.6(b), 5.6(d), 5.6(f), 5.6(h) and 5.6(j) display the correlation matrices of one set of simulated rainfall fields. As shown in the provided figures, the overall spatial dependence structure is reasonably preserved in the simulated fields using the v-copula model. In order to quantify the error, the values of the Mean Absolute Error (MAE) of the simulated and observed correlation matrices are given in Table 5.3. The table also provides the mean absolute error of the correlation matrices for the cases where fewer gauges were used for simulations (see Section 5.4 for details on the selected gauges). The the table indicates that the MAE increases as the number of rain gauges decreases. Additionally, a comprehensive comparison of the models' (Gaussian, t-copula, and v-copula) spatial dependence is provided in Chapter 8.

As discussed in Chapter 4, the temporal autocorrelations of radar rainfall error are reported to be rather low (Habib et al. (2008)). This

Table 5.3 The mean absolute error (MAE) of the correlation matrices of the simulated rainfall fields with respect to the observed (v-copula model).

Event ID	All Gauges	50% LW Gauges 50% GC Gauges	12% LW Gauges 25% GC Gauges
LW1-IV	0.051	0.090	0.178
LW2-IV	0.039	0.099	0.174
LW3-IV	0.031	0.085	0.201
LW4-IV	0.041	0.059	0.126
LW5-IV	0.063	0.112	0.219
GC1-II	0.059	0.133	0.130
GC2-II	0.041	0.082	0.099
GC3-II	0.039	0.089	0.148
GC4-II	0.038	0.082	0.131
GC5-II	0.016	0.090	0.148

issue has been confirmed for the events used in this study (see Figure 4.13). In the presented v-copula model, the temporal autocorrelation of error is not explicitly accounted for. However, by imposing error fields over radar estimates, the underlying temporal autocorrelations of radar estimates will be carried forward to the simulated fields. For the Little Washita events, Figures 5.7(a), 5.7(c), 5.7(e), 5.7(g) and 5.7(i) plot the temporal autocorrelations of the radar estimates and 500 simulated rainfall fields using the v-copula for the square-marked pixel shown in Figure 3.1. Similarly, for the Goodwin Creek events, Figures 5.7(b), 5.7(d), 5.7(f), 5.7(h) and 5.7(j) display the temporal autocorrelations of the radar estimates and 500 simulated rainfall fields for the square-marked pixel shown in Figure 3.2. Overall, the figures show that the simulated rainfall fields have quite similar temporal autocorrelation characteristics. It is noted, however, that the simulated ensembles exhibit slightly lower temporal autocorrelations due to disturbing radar rainfall data with random error fields.

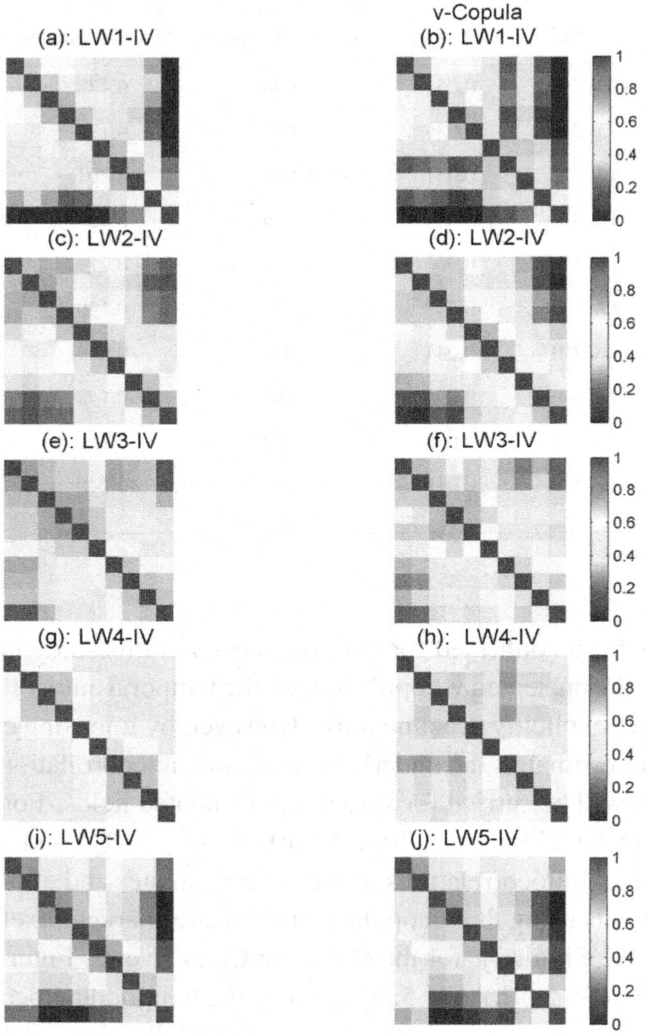

Figure 5.5 Rank correlation matrices of Events LW1-IV to LW5-IV (left); Rank correlation matrices of one set of simulated data using v-copula (right).

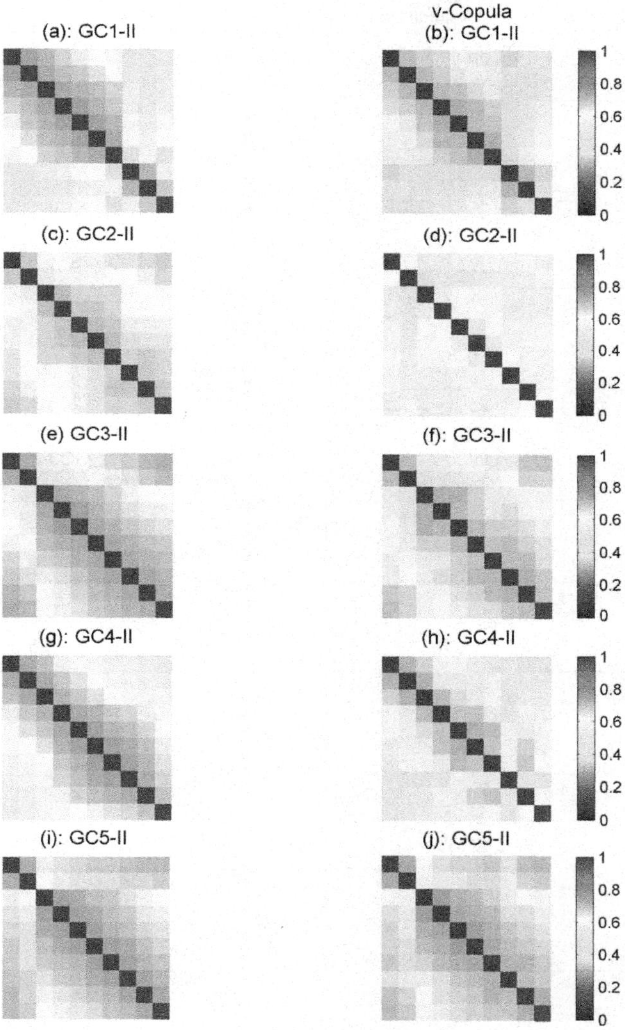

Figure 5.6 Rank correlation matrices of Events GC1-IV to GC5-IV (left); Rank correlation matrices of one set of simulated data using v-copula (right).

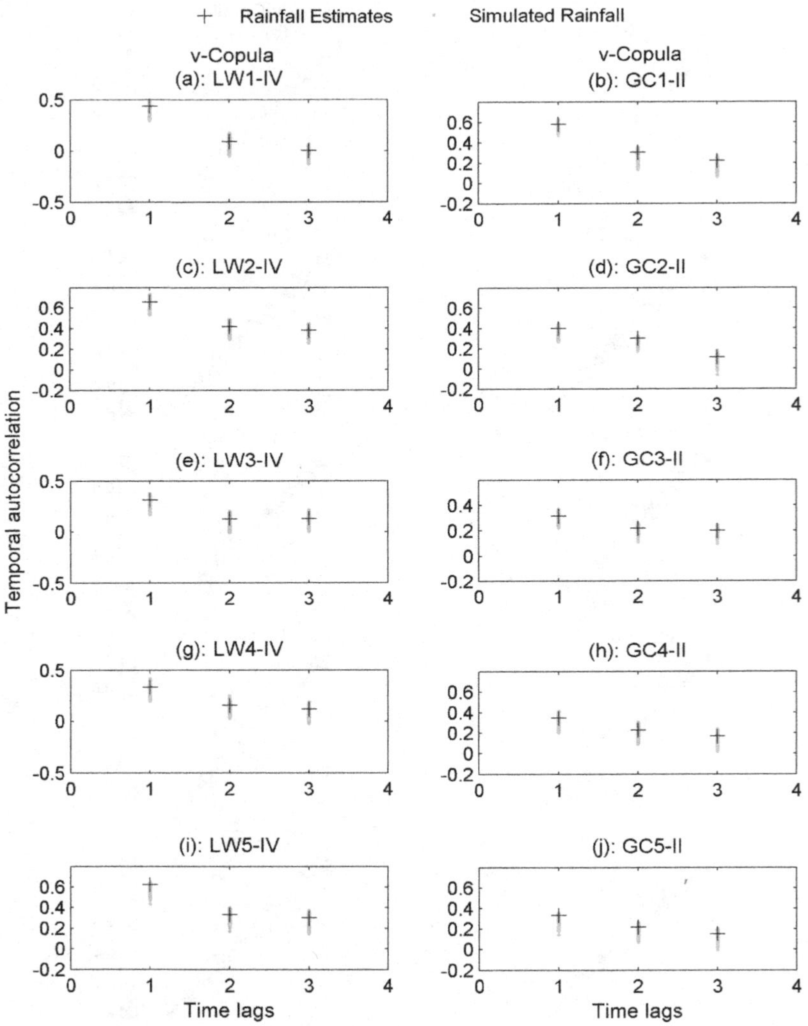

Figure 5.7 Temporal autocorrelations of rainfall estimates and 500 simulated rainfall realizations for: (left) Events LW1-IV to LW5-IV; (right) Events GC1-II to GC5-II.

5.6 EXTREME VALUES

A major difference associated with different copula families is the inherent tail dependence. A particular type of copula may provide strong upper tail dependence, whereas another copula may represent strong lower tail dependence. The selected copula family for a practical application has to be tested with respect to the extremes. In the following, the v-copula model is tested for probability occurrence of extreme values. In this study, the 90th percentile and 75th percentile of the radar estimates are assumed as the extreme value thresholds. In order to investigate the v-copula model with respect to extremes, the number of extreme occurrences above the thresholds in the simulated fields are compared with the radar rainfall estimates. For the Little Washita events, Figures 5.8(a) to 5.8(e) display the sum of the number of occurrences above the 90th percentile threshold (solid black lines), the mean of the number of occurrences above the 90th percentile in 500 realizations simulated using the v-copula (dashed lines) and the number of occurrences above the threshold of radar estimates in each simulated realization (gray lines).

Figures 5.8(f) to 5.8(j) present similar figures for the Goodwin Creek rainfall. In the figures, the x-axis shows the number of simulated realizations (here 500) and the y-axis represents the number of occurrences above the threshold (90th percentile) of the observations. One can see that in most cases the v-copula model slightly over estimates the number extreme occurrences, which indicates that the v-copula shows a positive tail dependence (see Chapter 2). For the 75th percentile, Figures 5.9(a) to 5.9(e) and Figures 5.9(f) to 5.9(j) plot the number of extreme occurrences above the threshold. Table 5.4 lists the error (%) in the number of extreme occurrences in the simulated rainfall ensembles with respect to the observed rainfall fields. The second and third columns show the error (%) for the 90th and 75th percentile thresholds, respectively.

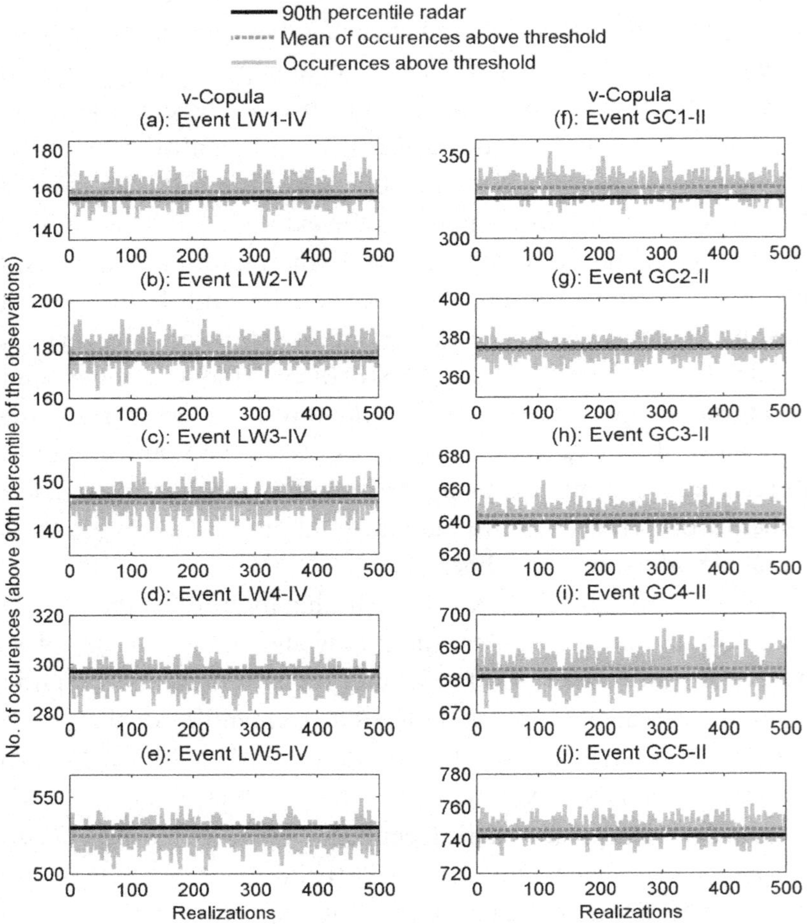

Figure 5.8 Total number of occurrences above the 90th percentile of radar data (solid black lines), mean number of occurrences above the same threshold in 500 realizations (dashed lines) and the number of occurrences above the 90th percentile of radar estimates in each simulated realizations.

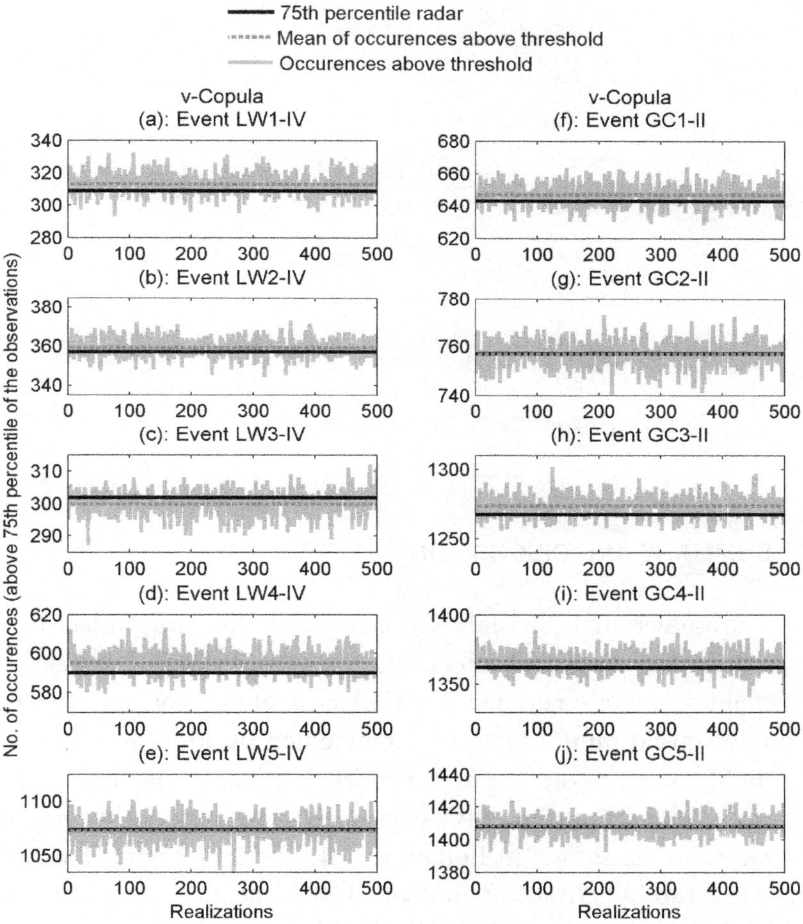

Figure 5.9 Total number of occurrences above the 75th percentile of radar data (solid black lines), mean number of occurrences above the same threshold in 500 realizations (dashed lines) and the number of occurrences above the 75th percentile of radar estimates in each simulated realizations.

Table 5.4 The error (%) in the number of extreme occurrences in the simulated fields (v-copula) with respect to the observed rainfall data.

Event ID	90th percentile	75th percentile
LW1-IV	2.1	1.2
LW2-IV	2.0	0.6
LW3-IV	2.1	1.0
LW4-IV	1.8	0.7
LW5-IV	1.2	0.4
GC1-II	1.9	0.8
GC2-II	0.7	0.1
GC3-II	0.8	0.5
GC4-II	0.8	0.4
GC5-II	0.7	0.1

5.7 SUMMARY AND DISCUSSION

It is well known that remotely sensed rainfall data are subject to different types of errors that affect the quality of rainfall estimates. In this chapter, a non-Gaussian copula-based model for simulation of remotely sensed rainfall data is implemented by imposing simulated error fields over remotely sensed rainfall estimates was presented. In this model, rainfall uncertainty is assumed to be a spatially correlated additive term, as shown in Equation 4.1. Several studies have confirmed that rainfall error is not independent of the magnitude of the rain rate (AghaKouchak et al. (2010b); Villarini et al. (2009); Ciach et al. (2007)). This property of rainfall uncertainty is accounted for by using a multiplicative error term ($R_i \times \epsilon$). Using this method, one can avoid unrealistically large error values where the rain rate is not significant.

In this model, a v-transformed copula was employed to describe the dependence structure of the rainfall data without the influence of the marginal distribution. Spatial dependencies of the simulated radar rainfall fields were then investigated by calculating their rank correlation matrix. The results revealed that this copula-based model preserves the spatial dependence of the simulated rainfall fields. It should be noted that no explicit accounting for error temporal autocorrelation was included in the model. However, owing to the inherent dominance of the temporal autocorrelation values, the results show that the trend of temporal autocorrelations in simulated fields will be similar to the temporal autocorrelation of the rainfall estimates.

It is noted that the issue of accounting for spatial dependencies using copulas requires further investigations, as the choice of copula itself plays an important role (Melchiori (2003); Venter (2002); Embrechts et al. (2001)). Using cross validation by taking the mean absolute error (MAE) as the estimator, the fitted copula was tested based on the available observations. Table 5.5 summarizes the results of the repeated random sub-sampling cross validation (Picard and Cook (1997)) for the fitted v-copula to the observations. Similar to the analysis presented in Chapter 4 for the case where only 5 reference gauges are used in simulations, the leave-one-out cross-validation method (LOOCV) is employed due to the limited observations (see column 4 in Table 5.5). In this method, a single observation from the original sample is put aside for validation, and the remaining observations are used as the training data. This is repeated for the number of observations such that each observation is used as the validation data once. It is noted that the mean absolute error does not change considerably as the number of gauges is reduced. Overall, it seemed that the v-transformed copula is an appropriate choice for modeling spatial dependencies. However, the parameter estimation and simulation process is computationally demanding.

In most rainfall simulation models, a standard distribution function is fit to the data and used for simulation in order to simplify the process. Often, the normal distribution is used for its simplicity in terms of parameter estimation and simulation of multivariate fields. In this work, however, the empirical marginal distribution of the observed rainfall error is applied so that simulated error values will have the same distribution function as that of the observed.

Table 5.5 The results of cross validation for the v-copula model (MAE estimator).

Event ID	All Gauges	50% LW Gauges 50% GC Gauges	12% LW Gauges 25% GC Gauges
LW1-IV	0.90	0.91	2.40
LW2-IV	0.72	1.14	2.44
LW3-IV	1.06	1.56	2.66
LW4-IV	0.98	1.08	2.35
LW5-IV	1.08	1.71	2.50
GC1-II	1.11	1.09	2.72
GC2-II	0.99	1.81	2.80
GC3-II	1.00	1.76	2.88
GC4-II	0.84	1.79	2.92
GC5-II	1.14	2.02	2.60

The v-copula model is also validated by generating rainfall ensembles with different numbers of gauges to investigate the estimated uncertainty when fewer gauges were available. The number of gauges used for simulations were reduced to 20 and 5 gauges for the Little Washita events and 10 and 5 gauges for the Goodwin Creek events. The results showed that the estimated uncertainties associated with the radar data enclosed the rain gauge measurements except at few time steps. It is pointed out that as the number of gauges decreases, the sample size of the observed error decreases. This may, but not necessarily, result in a shorter range of error (the difference between minimum and maximum error). One can intuitively conclude that any change in the error range and thus the empirical distribution of error will also affect the estimated uncertainty.

It is noted that in most radar pixels, one individual gauge is used to represent the true area average rainfall over a pixel size of (16 km^2/ 1 km^2), which may not be very accurate. However, based on the available rain gauge stations, this was the best possible approximation of the true area average rainfall values. Obviously, one can argue that unreliable rain gauge measurements may result in inaccurate true area average rainfall representation and thus cause errors in parameter estimation. Consequently, this will affect the simulated rainfall fields, especially in the case of conditional simulation. Overall, the results presented here indicate that using this model, one can generate multiple realizations of rainfall fields through the simulation of error fields conditioned on observed error values. A more detailed discussion is provided in Chapter 8.

CHAPTER 6

A RANDOM ERROR MODEL

6.1 INTRODUCTION

Characterization of radar-based rainfall uncertainties and their effects on hydrologic simulations is a challenging issue. The superposition of random error of different sources is one of the main factors causing the uncertainty of radar estimates. One way to express these uncertainties is to stochastically generate random error fields and impose them on radar measurements in order to obtain an ensemble of radar rainfall estimates. In Chapters 4 and 5, three copula-based models were presented for simulation of radar rainfall fields. In this chapter, a rainfall simulation model is introduced in which radar estimates are perturbed with a purely random error component and an error component that depends on the magnitude of rainfall rates. The presented model is parameterized with two uncertainty parameters and two other parameters that relate reflectivity values to rainfall rates (the exponent and multiplicative factors in the $Z - R$ relationship). The model parameters are estimated using the maximum likelihood method in order to account for heteroscedasticity in radar rainfall error estimates. Heteroscedasticity, alternatively known as variance heterogeneity, occurs when the distribution of residuals is dependent on the indicator variable (Petersen-Ø verleir (2004)). The model is implemented over the study areas, introduced in Chapter 3, to demonstrate the model performance. In order to obtain estimates of the radar rainfall error across the study areas, reference surface rainfall data are obtained from high resolution rain gauge measurements over the selected watersheds. The differences between reference surface rainfall data (R_{gauge}) and radar estimates (R_{radar}) are considered and termed as observed error. In the provided examples, ensembles of radar rainfall data are simulated by perturbing the radar estimates with generated radar rainfall error fields.

6.2 MODEL DESCRIPTION

It is well-known that radar measurements are associated with different error types (Wilson and Brandes (1979); Austin (1987); Hunter (1996); Pegram and Clothier (2001); Rico-Ramirez et al. (2007)) that can be grouped into three main classes: (1) physical biases; (2) inaccuracies in Z-R relationship; and (3) random error (Jordan et al. (2003)). Physical biases such as ground clutter, beam broadening and beam blockage can be removed using different algorithms as described in Steiner and Smith (2002), Venkatesan et al. (1997) and Dixon et al. (2005). Random error and measurement biases, however, are the most difficult types of radar error to eliminate and will remain even after removing the bias with an appropriate $Z - R$ function. The superposition of random errors from different sources is one of the main factors causing the uncertainty of radar estimates. Ciach et al. (2007) and Villarini et al. (2009) suggested that the systematic error has to be removed before any analysis is performed on the radar random error component. Similarly, the model presented here does not intend to capture all types of radar error, especially those related to physical issues (e.g. beam blockage) that may result in no rain at parts of a radar image.

One way to simulate multiple realizations of radar rainfall fields is to perturb radar estimates with random error fields. Recent studies (e.g., Ciach et al. (2007); Habib et al. (2008)), indicate that the radar error may depend on the magnitude of the rain rate. In order to account for such dependence, the following formulation is suggested:

$$R_i = \left(\frac{Z_i}{A}\right)^{1/b} + \left(\frac{Z_i}{A}\right)^{1/b} \epsilon_1 + \epsilon_2 \tag{6.1}$$

where : R_i = rain rate
Z_i = Reflectivity
ϵ_1 = proportion error
ϵ_2 = purely random error
A = multiplicative factor in $Z - R$ relationship
b = exponent in $Z - R$ relationship

For multi-sensor precipitation estimates that are available in rain rates (and not reflectivity), Equation 6.1 is expressed as:

$$R_i = P_i + P_i \times \epsilon_1 + \epsilon_2 \tag{6.2}$$

where : P_i = multi-sensor precipitation estimates (rain rate)

In this formulation, R represents reference surface rainfall measurements which are typically obtained from rain gauge observations. The term, $(Z_i/A)^{1/b}\,\epsilon_1$, represents the error component that is dependent on the magnitude of the reflectivity or rain rate, whereas ϵ_2 accounts for purely random error sources. Assuming normally distributed error, the parameters of ϵ_1 and ϵ_2 are the mean and standard deviation. Ciach et al. (2007) showed that unbiased radar rainfall random error can be reasonably described by a normal distribution. Since the overall radar error is a sum of errors from different sources, this assumption of a Gaussian distribution of error is reasonable, as previously explained. Based on the central limit theorem, the sum of all errors will tend toward a Gaussian distribution, regardless of the distribution of each individual source of error. The normality assumption, however, is tested in the following sections. In general, the model structure allows for the selection of an appropriate probability distribution for radar error (not necessarily Gaussian) based on available data. However, using the likelihood function for parameter estimation may become computationally too expensive. Note that the aim of this study is to present a simple model that performs reasonably well with minimal computational costs, even though this may result in a imperfect agreement in some cases.

The error terms in Equation 6.1 can be summed up to one error term representing the total error ($\epsilon = (Z_i/A)^{1/b}\,\epsilon_1 + \epsilon_2$):

$$R_i - \left(\frac{Z_i}{A}\right)^{1/b} \rightarrow \epsilon\left(0, \sigma^2\right) \tag{6.3}$$

where σ in Equation 6.3 is:

$$\sigma = \sqrt{\left(std\left(\frac{Z_i}{A}\right)^{1/b} \cdot \sigma_1\right)^2 + \sigma_2^2} \tag{6.4}$$

The term σ is the standard deviation of the total error (both error terms) in Equation 6.1. Substituting the probability density function of the normal distribution (Equation 6.5) into Equation 6.1 with some algebraic manipulation yields the following likelihood function (see Equation 6.6):

$$\epsilon(Y_1, ..., Y_n|0, \sigma^2) = \left(\frac{1}{2\pi\sigma^2}\right)^{\frac{n}{2}} exp\left(-\frac{\sum_n^{i=1}(Y_i)^2}{2\sigma^2}\right) \tag{6.5}$$

$$\text{where}: \quad Y_i = R_i - \left(\tfrac{Z_i}{A}\right)^{1/b}$$

$$L = -\frac{n}{2}\ln 2\pi - \frac{n}{2}\ln\left(\left(std\left(\frac{Z_i}{A}\right)^{1/b} \cdot \sigma_1\right)^2 + \sigma_2^2\right)$$

$$-\frac{1}{2}\frac{\sum_{i=1}^{n}\left(R_i-\left(\frac{Z_i}{A}\right)^{1/b}\right)^2}{\left(std\left(\frac{Z_i}{A}\right)^{1/b}\cdot\sigma_1\right)^2+\sigma_2^2} \tag{6.6}$$

where : σ_1 = standard deviation of the proportion error
σ_2 = standard deviation of the random error

In Equation 6.5, the term Y refers to the total error associated with radar data defined as rain gauge measurements (R_i) minus radar estimates ($(Z_i/A)^{1/b}$). When multi-sensor rainfall estimates are used as input for simulations, instead of Equation 6.6, the likelihood function is expressed as:

$$\begin{aligned} L &= -\frac{n}{2}\ln 2\pi - \frac{n}{2}\ln\left((std \times P_i \cdot \sigma_1)^2+\sigma_2^2\right) - \\ &\quad \frac{1}{2}\frac{\sum_{i=1}^{n}(R_i-P_i)^2}{(std \times P_i \cdot \sigma_1)^2+\sigma_2^2} \end{aligned} \tag{6.7}$$

6.3 PARAMETER ESTIMATION

For radar reflectivity data, the parameters b, A, σ_1^2 and σ_2^2 are to be estimated simultaneously using the maximum likelihood method whereby the parameters are determined such that the likelihood function (probability of the sample data) is maximized. Using this approach, reflectivity is transformed into rain rate with parameters that describe the uncertainty of radar estimates. From a practical viewpoint, the method of maximum likelihood has proven robust and reliable in a number of hydrologic applications (see Sorooshian et al. (1983); Haddad et al. (2007) and Montopoli and Marzano (2007)). For details on the parameter estimation techniques using the maximum likelihood, readers are referred to Brandt (1999). It is noted that for multi-sensor rainfall estimates (Little Washita rainfall events), the number of parameters reduces to two standard deviations (σ_1^2 and σ_2^2) of the proportion error (ϵ_1) and random error components (ϵ_2) in Equation 6.2.

For all rainfall events used in this study, the model parameters are estimated and listed in Table 6.1. Note that the parameter estimation procedure must be controlled to make sure that the bias of the estimated rain rates is negligible. This result can be achieved by allowing for a minimal bias ($\pm 2\%$) and estimating the parameters such that the likelihood function is maximized and the overall bias is kept within the minimally allowed bias value. Note that the standard deviation of the total error can be obtained using Equation 6.4.

Table 6.1 Model parameters estimated for the random error model.

Event	σ_1^2	σ_2^2
LW1-IV	0.28	0.31
LW2-IV	0.29	0.34
LW3-IV	0.27	0.28
LW4-IV	0.30	0.32
LW5-IV	0.34	0.33
GC1-II	0.33	0.35
GC2-II	0.35	0.26
GC3-II	0.45	0.36
GC4-II	0.34	0.43
GC5-II	0.36	0.25

6.4 SIMULATION OF RADAR ERROR FIELDS

Using stochastic simulation techniques, one can generate multiple equally probable (likely) realizations with similar statistical properties. The process is termed as conditional simulation when generated fields are conditioned on a set of observed (input) data. In the present study, radar error fields are simulated conditioned on observed error values, which are estimated by comparing rain gauge measurements and radar estimates at rain gauge locations.

The simulation procedure used in this study can be summarized in 4 steps. First, an $m \times n$ normally distributed field termed as ϵ_2 is simulated using the model parameter σ_2, where m and n are the dimensions of the reflectivity field. In the second step, a multivariate normal field (ϵ_1) is simulated using the model parameter σ_1 and multiplied by the radar estimates. In order to condition on observed error values at rain gauge locations, a random value from a normal distribution with a standard deviation of σ_1 is selected such that the sum of both error terms is similar to the observed error value. At pixels where no rain gauges exist and thus no observed error values are available, ϵ_1 is randomly selected from the distribution. In the third step, both simulated error fields are imposed on the radar estimates in order to obtain one realization of simulated radar estimates. In the fourth and final step, the first through third steps are repeated to obtain an ensemble of radar estimates.

It is worth pointing out that in this model (E-Model hereafter), the proportion error (ϵ_1) is included to account for the proportionality of rainfall error to the magnitude of the rain rate (similar to the approach explained in Chapter 4). For example, Figures 6.1(a) and (b), respectively, plot an example of simulated proportion error ($(Z_i/A)^{1/b}\,\epsilon_1$) and random error ($\epsilon_2$) versus the rain gauge measurements for Event GC3-II. As shown, ϵ_1 is proportional to the magnitude of rain rate, whereas random error (ϵ_2) remains independent. Figure 6.1(c) presents the total error which is the sum of both error terms.

Following the simulation steps described above, ensembles of radar rainfall estimates are generated for the Little Washita and Goodwin Creek events based on Equation 6.1. It is worth remembering that in the presented error model (E-Model), the radar error fields are simulated based on the multivariate normal distribution, and unlike the previous models, the marginal of the observed error is NOT applied to the simulated error fields. In the instance of LW1-IV, Figure 6.2(a) and Figure 6.2(b) display a radar rainfall field and rain gauge measurements corresponding to the same time step. Figure 6.2(c) presents an example of conditionally simulated rainfall image using the E-Model

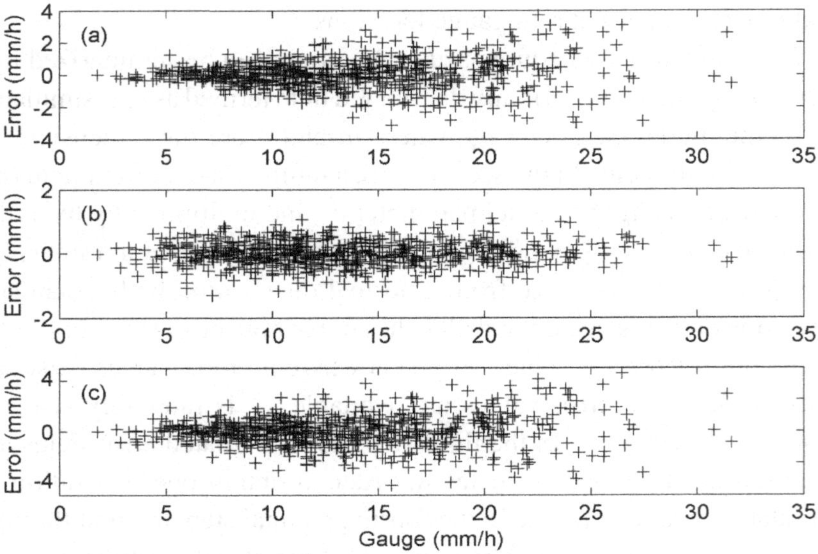

Figure 6.1 (a) Simulated proportion error $((Z_i/A)^{1/b}\,\epsilon_1)$; (b) simulated random error (ϵ_2); (c) total error versus the rain gauge measurements.

after perturbing radar estimates with simulated error fields. Figures 6.2(d) to 6.2(o), provide similar figures for rainfall Events LW2-IV to LW5-IV. As shown in the presented figures, the E-Model, honors the rain gauge measurements (Figures 6.2(b), 6.2(e), 6.2(h), 6.2(k) and 6.2(n)) at their locations (see Figures 6.2(c), 6.2(f), 6.2(i), 6.2(l), 6.2(o)). For the Goodwin Creek events, on the other hand, Figures 6.3(a) to 6.3(o) demonstrate examples of radar estimates, gauge measurements and simulated rainfall fields.

The presented error model is employed to obtain 500 equally likely realizations to describe the uncertainty of rainfall estimates. For the Little Washita events (LW1-IV, LW2-IV, LW3-IV, LW4-IV and LW5-IV), Figures 6.4(a), 6.4(c), 6.4(e), 6.4(g) and 6.4(i) plot radar estimates and simulated rainfall ensembles (500 realizations) over one radar pixel (the square-marked pixel shown in Figure 3.1). For the Goodwin Creek events (GC1-II, GC2-II, GC3-II, GC4-II and GC5-II),

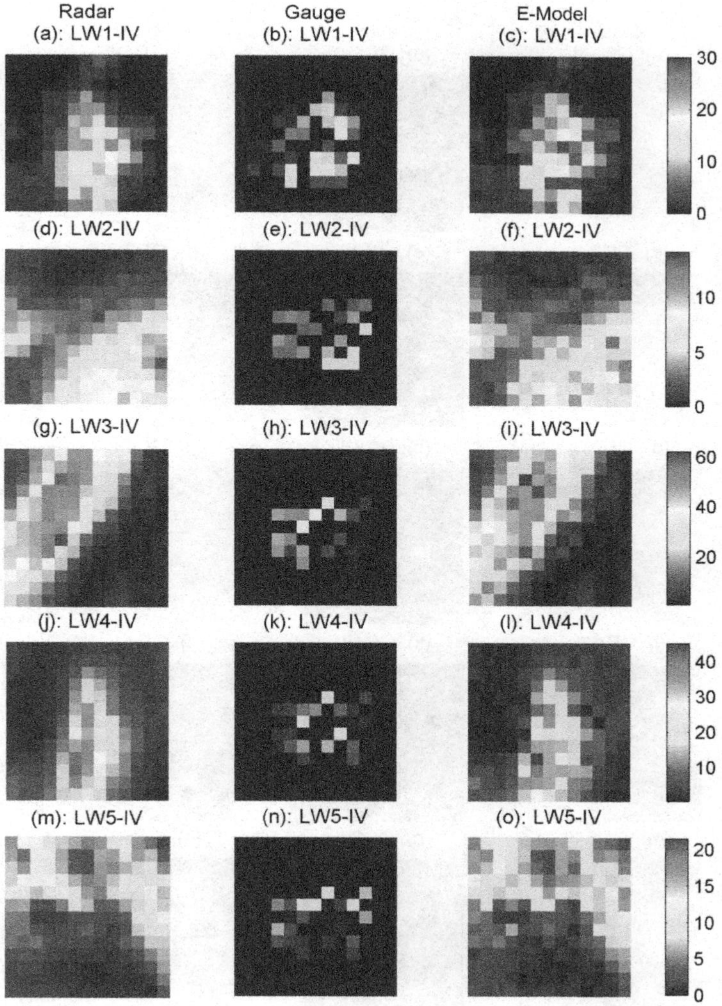

Figure 6.2 Event LW1-IV; (a): radar-based rainfall estimates. (b): rain gauge measurements. (c): simulated rainfall using the error model (E-Model). Figures (d) to (f), (g) to (i), (j) to (l) and (m) to (o), display similar plots for Events LW2-IV, LW3-IV, LW4-IV and LW5-IV, respectively.

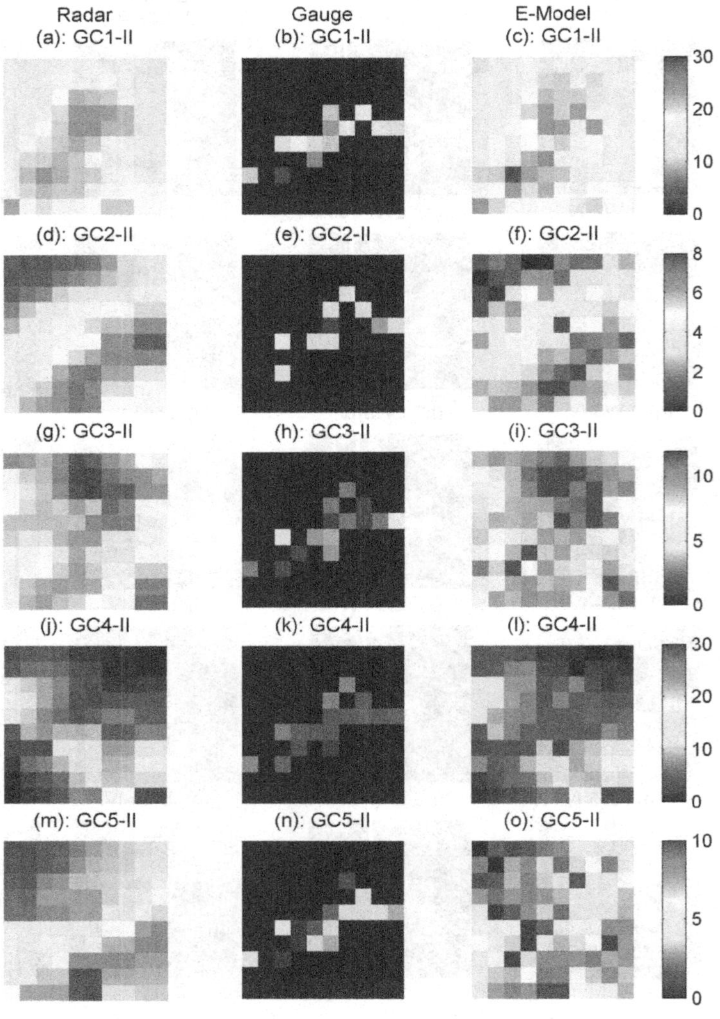

Figure 6.3 Event GC1-II; (a): radar-based rainfall estimates. (b): rain gauge measurements. (c): simulated rainfall using the error model (E-Model). Figures (d) to (f), (g) to (i), (j) to (l) and (m) to (o), display similar plots for Events GC2-II, GC3-II, GC4-II and GC5-II, respectively.

Figures 6.4(b), 6.4(d), 6.4(f), 6.4(h) and 6.4(j), display simulated rainfall ensembles over the square-marked pixel shown in Figure 3.2. In these figures, the solid black and gray lines represent radar estimates and simulated rainfall realizations, respectively. It is noted that in this model the proportionality of error to the magnitude of rain rates is accounted for using the term ϵ_1 in Equation 6.1.

The E-Model is verified through simulation and parameter estimation based on different numbers of rain gauges. In the instance Event LW1-IV, for example, the model is used for simulation based on 20 rain gauges (50%) and 5 rain gauges (12.5%). Figures 6.5(a), 6.5(b) and 6.5(c) display rainfall ensembles simulated using the v-copula based model for the x-marked pixel (shown in Figure 3.1), which contains one of the removed rain gauges. In Figure 6.5(a) 41 rain gauges are used for simulation, whereas in Figures 6.5(b) and 6.5(c), 20 and 5 rain gauges are included for ensemble generation. Rainfall ensembles are also generated for Events GC1-II, GC2-II, GC3-II, GC4-II and GC5-II using 10 rain gauges (50%) and 5 rain gauges (25%). For rainfall Event GC3-II, Figures 6.5(d), 6.5(e) and 6.5(f) display the generated ensembles over the x-marked pixel, shown in Figure 3.2, based on different numbers of gauges. In Figures 6.5(d) and 6.5(e), 18 rain gauges and 10 rain gauges, marked with triangles and diamonds in Figure 4.9(b), are used in parameter estimation and simulations, whereas in Figure 6.5(f), 5 gauges (marked with diamonds in Figure 4.9(b)) are included. In the presented figures, the solid lines represent radar estimates and the dashed lines represent rain gauge measurements. The gray lines are the estimated rainfall uncertainty over the entire storm. As expected from a reasonable ensemble, the ground reference measurements are consistently enclosed, with the exception of a few time steps.

In order to numerically evaluate how well the estimated uncertainty encompasses rain gauge measurements, the number of time steps (n_{out}) where the ground reference measurements fall outside the generated ensembles is counted. Table 6.2 gives the n_{out} values in percentage for each of the rainfall events used in this study. The table indicates that reducing the number of gauges does not result in dramatic changes in the n_{out} values. Furthermore, in this model, the n_{out} values are

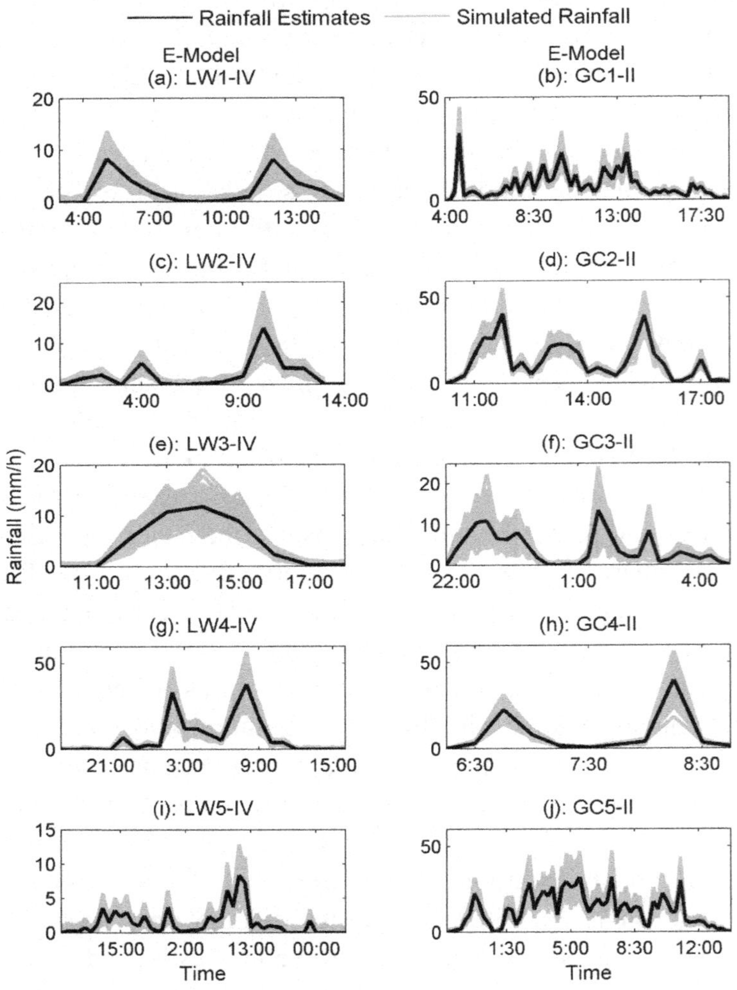

Figure 6.4 Simulated rainfall ensembles (500 realizations) using the E-Model for: the Little Washita events ((a),(c),(e),(g),(i)); the Goodwin Creek events ((b),(d),(f),(h),(j)).

less than previous models. The reason is that in previous models the empirical CDF of error is applied to the simulated error values, and thus they are within the range of observed error values. In this model, however, the error is assumed to be normally distributed and the simulated values may exceed the range of observed error values. It is noted that in this model the estimated uncertainty is described by parameters (σ_1 and σ_2) that depend on the agreement between radar estimates and gauge data. In fact, large uncertainty parameters and thus, high variability in simulated fields is expected when the deviation of rain gauge measurements from radar estimates is significant and vice versa. For more detailed comparisons between the models, the reader is referred to Chapter 8.

Table 6.2 The number of time steps, n_{out} (%), where the estimated uncertainty using the E-Model did not enclose the rain gauge measurements.

	All Gauges	50% LW Gauges 50% GC Gauges	12% LW Gauges 25% GC Gauges
LW1-IV	0.2	2.0	3.7
LW2-IV	0.5	0.9	4.0
LW3-IV	0.2	0.9	3.0
LW4-IV	0.7	1.4	3.4
LW5-IV	0.9	1.6	3.1
GC1-II	0.3	2.0	3.7
GC2-II	0.7	1.7	5.1
GC3-II	0.9	1.8	4.9
GC4-II	0.8	1.9	4.3
GC5-II	0.9	2.0	5.1

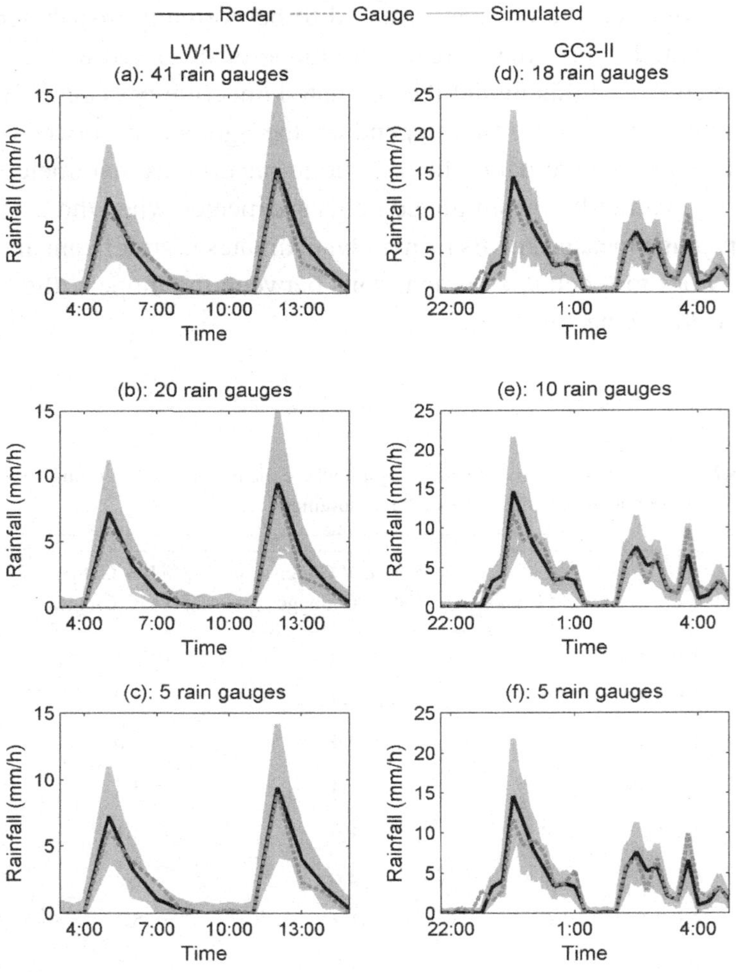

Figure 6.5 Simulated rainfall ensembles using the E-Model: ((a: 41 gauges), (b: 20 gauges; marked with triangles and diamonds in Figure 4.9(a)) and (c: 5 gauges; marked with diamonds in Figure 4.9(a))); ((d: 20 gauges), (e: 10 gauges; marked with triangles and diamonds in Figure 4.9(b)) and (f: 5 gauges; marked with diamonds in Figure 4.9(b)).

6.5 SPATIO-TEMPORAL DEPENDENCIES

In a recent study, Habib et al. (2008) showed that the spatial dependence of radar error is not negligible and can have a significant affect on streamflow hydrologic predictions. In the present study, the purely random error term (ϵ_2) is defined to be uncorrelated in space. However, with respect to the proportion error term, when ϵ_1 is multiplied by the radar estimates (Equation 6.1), the resulting error field ($(Z_i/A)^{1/b}\epsilon_1$) will be spatially correlated. This link in spatial correlation results from the underlying radar field. To demonstrate the model performance with respect to the spatial dependence structure, the correlation matrices of the observed and one set of simulated data are presented here. For the Little Washita events, Figures 6.6(a), 6.6(c), 6.6(e), 6.6(g) and 6.6(i) show the Spearman correlation matrices for the rainfall estimates of 10 MPE pixels, whereas Figures 6.6(b), 6.6(d), 6.6(f), 6.6(h) and 6.6(j) display the correlation matrices for one set of simulated rainfall fields obtained using the E-Model. For Goodwin Creek events, Figures 6.7(a), 6.7(c), 6.7(e), 6.7(g) and 6.7(i) present the Spearman correlation matrices for the rainfall estimates of 10 MPE pixels, while Figures 6.7(b), 6.7(d), 6.7(f), 6.7(h) and 6.7(j) demonstrate the correlation matrices of one set of simulated rainfall fields. Overall, one can see that the observed and simulated patterns are quite similar. However, the rank correlation values in simulated fields are less mainly due to the perturbation. The Mean Absolute Error (MAE) of the Spearman rank correlation (equation 4.5) matrices of the observed and simulated radar estimates are compared and listed in Table 6.3.

In this model, the error terms ϵ_1 and ϵ_2, are assumed to be temporally uncorrelated, as the temporal autocorrelations of the total error are found to be low at the first time lag and close to zero for larger time lags (Habib et al. (2008)). However, similar to the previous models, when ϵ_1 is multiplied by the radar estimates, the resulting error field ($(Z_i/A)^{1/b}\,\epsilon_1$) will carry some of the temporal self-correlation of the unperturbed radar field. For the rainfall Events LW1-IV to LW5-IV, Figures 6.8(a), 6.8(c), 6.8(e), 6.8(g) and 6.8(i) display the temporal self-correlations of the radar estimates and 500 simulated rainfall fields obtained using the E-Model for the square-marked pixel shown in Figure 3.1. For the rainfall events GC1-II to GC5-II, Figures 6.8(b), 6.8(d), 6.8(f), 6.8(h) and 6.8(j) plot the temporal self-correlations of the radar data and 500 generated rainfall ensembles for the square-marked pixel shown in Figure 3.2. The figures indicate the generated rainfall realizations have similar temporal autocorrelation characteristics.

Table 6.3 The mean absolute error (MAE) of the correlation matrices of the simulated rainfall fields with respect to the observed (E-Model).

Event ID	MAE
LW1-IV	0.101
LW2-IV	0.112
LW3-IV	0.098
LW4-IV	0.143
LW5-IV	0.126
GC1-II	0.132
GC2-II	0.203
GC3-II	0.217
GC4-II	0.179
GC5-II	0.226

6.6 EXTREME VALUES

The presented model is based on the Gaussian assumption for the radar rainfall error. As discussed in Chapter 2, the model is not expected to show asymptotic dependence. In previous chapters, the behavior of different copulas with respect to the simulation of simultaneous extremes was investigated. In this section, similar analysis is performed to evaluate the probability occurrence of multiple extreme values above certain thresholds. Assuming the 90th percentile and 75th percentile of

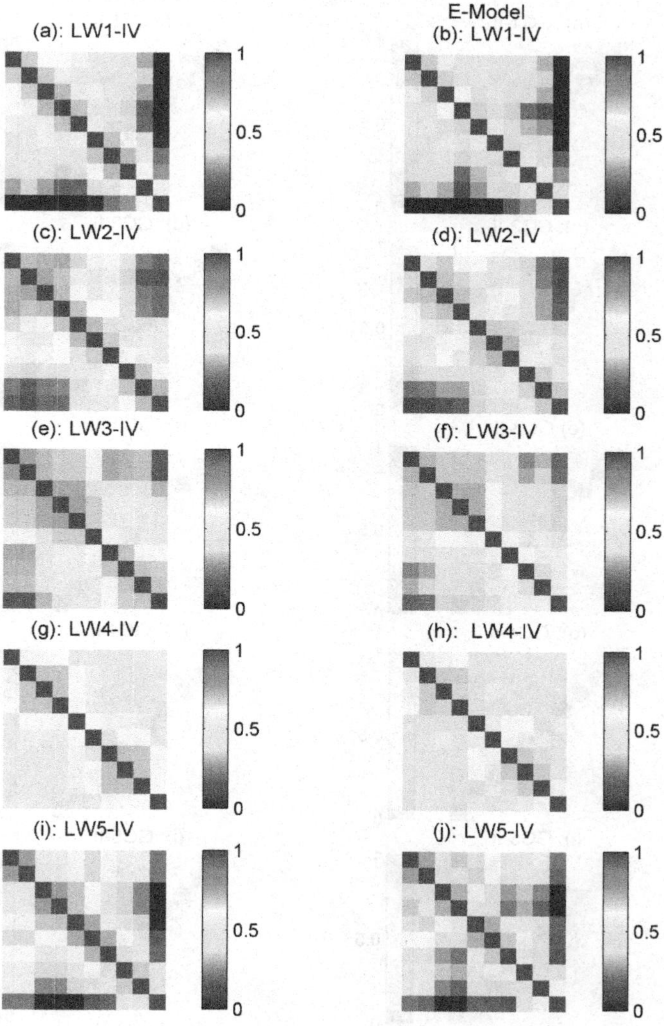

Figure 6.6 Rank correlation matrices of Events LW1-IV to LW5-IV (left); Rank correlation matrices of one set of simulated data using E-Model (right).

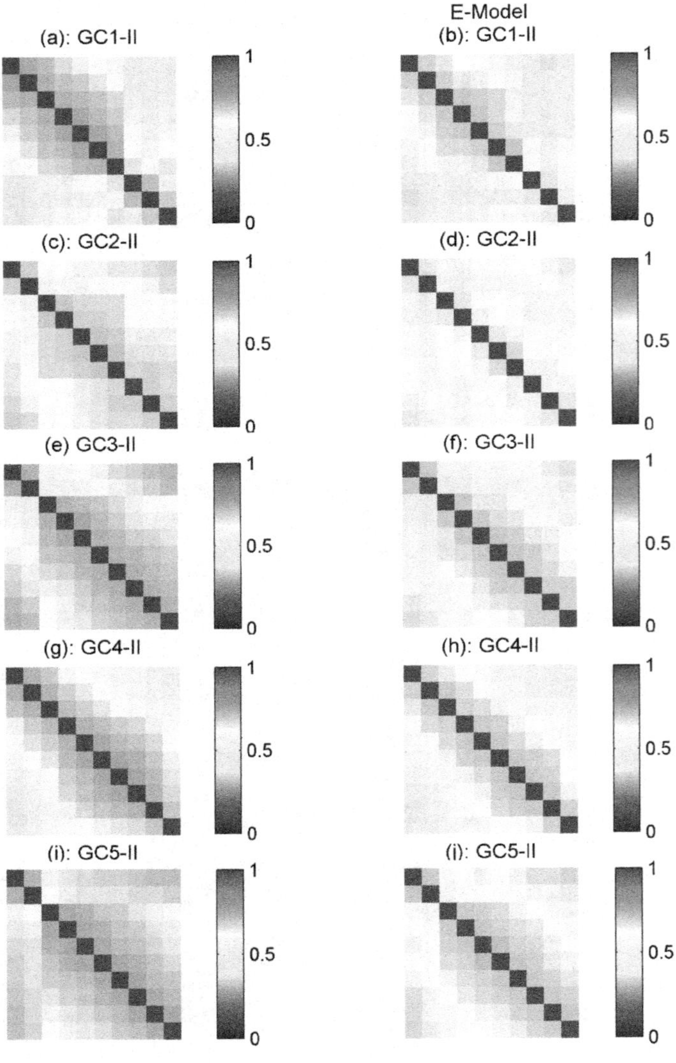

Figure 6.7 Rank correlation matrices of Events GC1-IV to GC5-IV (left); Rank correlation matrices of one set of simulated data using E-Model (right).

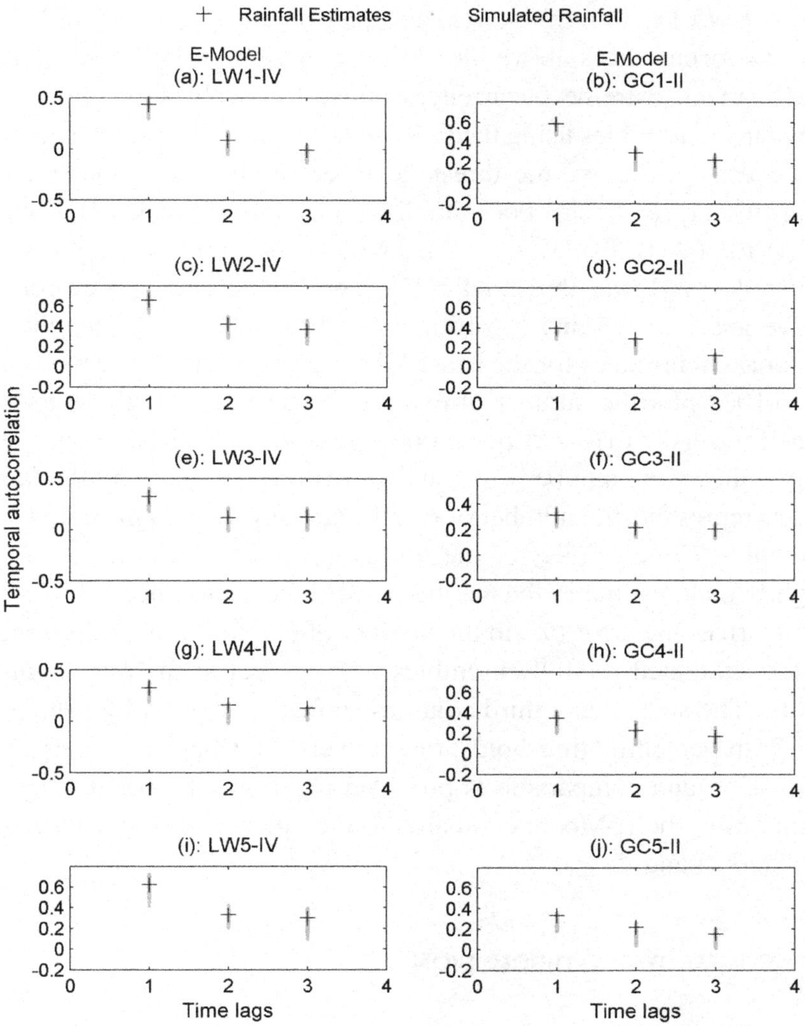

Figure 6.8 Temporal autocorrelations of rainfall estimates and 500 simulated rainfall realizations for: (left) Events LW1-IV to LW5-IV; (right) Events GC1-II to GC5-II.

the radar estimates as the threshold values, simultaneous occurrences of rainfall rates above the thresholds in the simulated rainfall values are compared with the radar estimates. For rainfall Events LW1-IV to LW5-IV, Figures 6.9(a) to 6.9(e) plot the total number of extreme occurrences above the 90th percentile threshold (solid black lines), mean extreme occurrences above the 90th percentile in the simulated ensembles using the E-Model (dashed lines) and the number of occurrences above the threshold in each simulated radar rainfall realization (gray lines). For rainfall Events GC1-II to GC5-II, Figures 6.9(f) to 6.9(j) display the number of occurrences above the same threshold. A lower threshold of 75th percentile is also used for the above analysis. Figures 6.10(a) to 6.10(e) present the number of extreme occurrences for the Little Washita events and Figures 6.10(f) to 6.10(j) plot the number of extreme occurrences for the Goodwin Creek watershed (for 75 percentile threshold). In these figures, the x-axis shows the number of simulated realizations (here 500) and the y-axis represents the number of occurrences above the threshold (90th percentile/75 percentile) of the observations. As shown, the model slightly underestimates the number of extreme occurrences. Table 6.4 summarizes the error (%) in the number of extreme value occurrences in the simulated rainfall ensembles with respect to the radar rainfall fields. The second and third columns show the error (%) for the 90th and 75th percentile thresholds, respectively. In Chapter 8, a detailed discussion and comparison is provided regarding the occurrences of extremes in the E-Model compared to the other models introduced in Chapters 4 and 5.

6.7 SUMMARY AND DISCUSSION

It is recognized that total radar rainfall estimation error is a result of different error sources that are superimposed on one another. The simulation model introduced in this chapter uses this concept to generate an ensemble of radar rainfall fields by perturbing radar estimates with conditionally simulated random error fields. The resulting realizations of radar rainfall fields can be used for assessing the impact

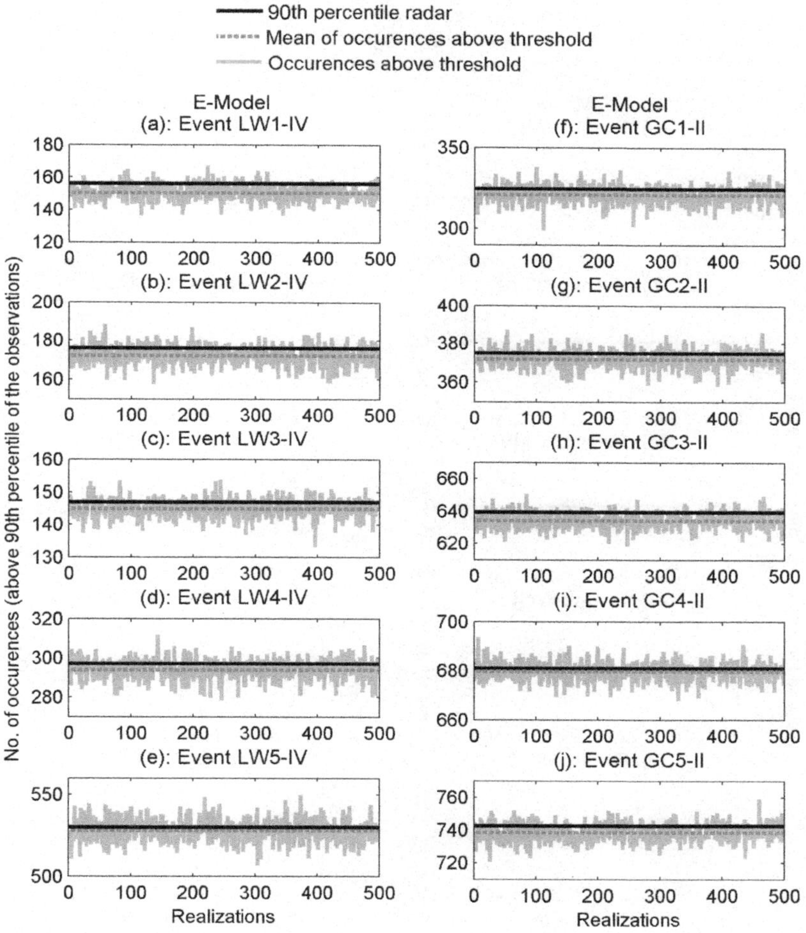

Figure 6.9 Total number of occurrences above the 90th percentile of radar data (solid black lines), mean number of occurrences above the same threshold in 500 realizations (dashed lines) and the number of occurrences above the 90th percentile of radar estimates in each simulated realizations.

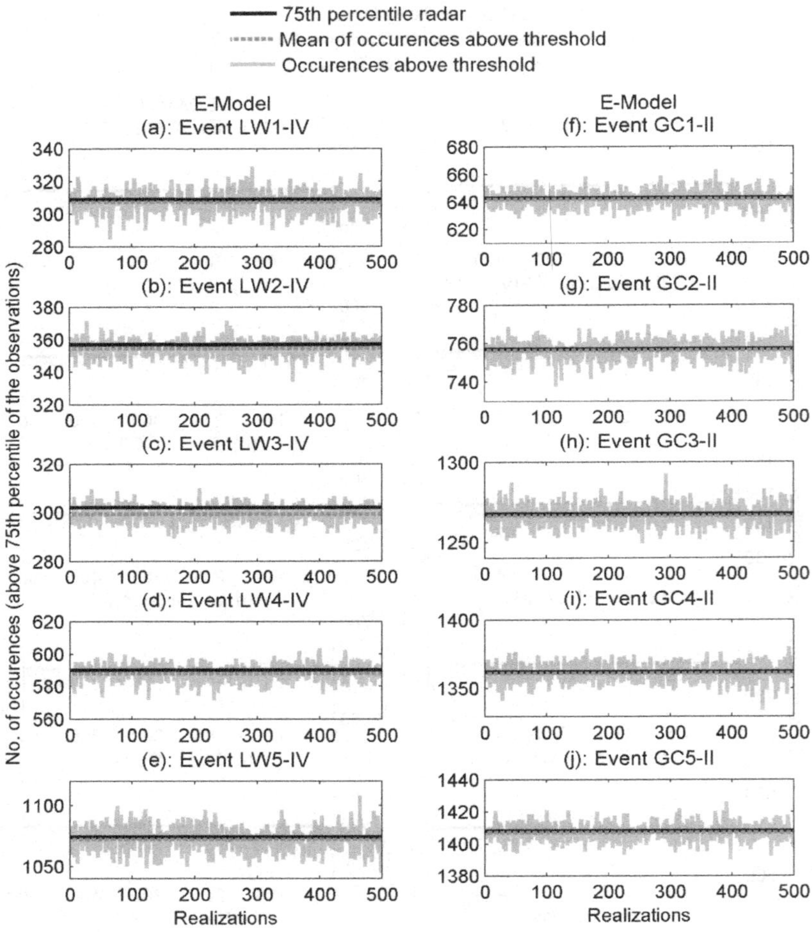

Figure 6.10 Total number of occurrences above the 75th percentile of radar data (solid black lines), mean number of occurrences above the same threshold in 500 realizations (dashed lines) and the number of occurrences above the 75th percentile of radar estimates in each simulated realizations.

Table 6.4 The error (%) in the number of extreme occurrences in the simulated fields (E-Model) with respect to the observed rainfall data.

Event ID	90th percentile	75th percentile
LW1-IV	3.1	1.7
LW2-IV	3.0	1.3
LW3-IV	2.1	0.9
LW4-IV	1.7	0.8
LW5-IV	1.2	0.8
GC1-II	1.7	0.9
GC2-II	1.2	0.8
GC3-II	1.0	0.6
GC4-II	1.1	0.6
GC5-II	1.2	0.7

of uncertainties associated with radar rainfall estimates on hydrologic predictions. In this model, radar uncertainty is described with two error components: purely random error and an error component dependent on the magnitude of the rain rate. Radar reflectivity fields or multi-sensor rain rate fields can be used as the input of the model to generate an ensemble of rainfall estimates. The model parameters are estimated using the maximum likelihood technique, assuming that the error terms are normally distributed. Application of this model for simulation of rainfall ensembles is tested to demonstrate the model performance. Using the presented model, multiple realizations of radar error fields are generated and used to perturb radar estimates. The results show that simulated realizations have similar statistical properties as those of unperturbed radar estimates and that the simulated ensemble encompasses ground reference measurements across the study area.

It should be noted that no explicit accounting for error autocorrelation is included in the model (both error terms, ϵ_1 and ϵ_2, are assumed to have no self-correlation in space and time). However, the self-correlation of the radar field will be carried forward when ϵ_1 is multiplied by the radar estimates, resulting in the proportion error term $((Z_i/A)^{1/b}\,\epsilon_1)$. Self-dependencies of the simulated radar rainfall fields are then investigated by calculating both the temporal autocorrelation coefficients at various temporal lags and the Spearman rank spatial correlation matrices. The results reveal that imposing radar error fields over unperturbed radar estimates by using the presented error model does not destroy the underlying spatial dependence structure. Furthermore, the temporal autocorrelations in the simulated fields are similar to that of the unperturbed radar field, which results from the dominance of the temporal autocorrelations of the unperturbed radar field.

Another simplification of the model is that the error components are assumed to be normally distributed. It is noted, however, that considerable deviation of radar error distribution from normality may affect the model performance. While several studies showed that radar error with temporal resolution of hourly and above can be reasonably modeled using a normal distribution (Ciach et al. (2007) and Villarini et al. (2009)), the distribution of high resolution radar error data may significantly deviate from the normal distribution. Theoretically, based on available data, any probability distribution function can be fit and assumed for radar error. However, the parameter estimation, based on likelihood approach, may require extensive computational effort and time. We emphasize that in this study, the goal is to present a simple model that performs reasonably well with minimal computational costs, considering that simplifying assumptions behind the model may result in a imperfect agreement in some cases.

CHAPTER 7

APPLICATION TO STREAMFLOW ANALYSIS

7.1 INTRODUCTION

There are various applications for simulated rainfall ensembles in the area of hydrology and meteorology. Using ensemble analysis, for example, one can evaluate flood prediction, uncertainty and risks associated with a given precipitation. This method uses many precipitation estimates instead of a single realization. Floods are the most life threatening hydrologic phenomena and result in extensive property damage and loss of life each year. Ensemble analyses can improve assessment of flood forecast uncertainty, which is influential in water resources management. Furthermore, climate change studies may also utilize simulated precipitation fields as future precipitation estimates in order to investigate long term changes in the hydrologic cycle of a specific region. Modeling climate changes with abnormal variations in one or more weather variables may allow exploration of subsequent affects on different hydrologic processes. For example, by projecting how precipitation may change over a significantly long period, one my be able to investigate regional changes in surface runoff and subsurface flow. Evaluating these impacts is important yet challenging, as surface runoff and groundwater availability are sensitive to changes in weather variables, particularly precipitation. Applications of rainfall ensembles are beyond the scope of this research; however, an example implication of simulated rainfall ensembles is presented in this chapter. This example utilizes a physically based hydrologic model (GSSHA) to implement rainfall ensembles obtained from the presented models in Chapters 4, 5 and 6.

The GSSHA (Gridded Surface Subsurface Hydrologic Analysis) model is selected mainly because it has been developed and calibrated for various hydrologic applications in the same study area (Habib et al. (2008)). Physically based models are meant to solve governing equations such as conservation of mass, momentum, flow and transport. In the following, a brief overview of the model is presented.

7.2 GSSHA MODEL

The GSSHA model is a fully distributed-parameter process-based hydrologic model developed by the US Army Corps of Engineers Engineering Research and Development Center (USACE ERDC). The model is intended to describe hydrologic processes and evaluate the surface runoff. The model is capable of simulating saturated source areas, surface runoff, overland water retention, infiltration, exfiltration, evapotranspiration, two-dimensional overland flow, groundwater discharge to streams, and one dimensional channel routing. The model allows simulation of the soil moisture profile in the unsaturated zone using different methods including Green and Ampt (Green and Ampt (1911)) and Richard's equation (Richards (1931)) as well as predicting different runoff generation mechanisms such as Hortonian (i.e., infiltration-excess) and non-Hortonian (i.e., saturation-excess) mechanisms (Chow et al. (1988)).

The model uses finite difference and finite volume methods to solve partial differential equations of mass conservation in order to guarantee an overall mass balance. The GSSHA model is an enhancement version of the CASC2D model (Ogden and Julien (2002)) which is used to estimate the surface runoff in arid to semi-arid basins. The GSSHA model benefits from all functionality of the CASC2D as well as the ability to simulate saturated and unsaturated groundwater flow. The model has been tested and verified on a number of different basins. The model was found to be efficient and to able to simulate runoff in both Hortonian and non-Hortonian basins with very little bias (Downer and Ogden (2002)). The capability of GSSHA to model the soil moistures and groundwater levels at the grid scale has also been verified (Downer and Ogden (2002)). In the following a brief overview of different processes and their multiple solution methods is provided:

- Infiltration module: Green and Ampt, multi-layered Green and Ampt, Green and Ampt with Redistribution, GAR, (Ogden and Saghafian (1997)), and Richard's Equation.
- Evapotranspiration: Deardorff (Deardorff (1997)), Penman-Monteith (Monteith (1981); Monteith (1981)) with seasonal canopy resistance.
- Channel routing: 1-D up-gradient explicit diffusive wave is used for channel routing.
- Overland flow routing: 2-D diffusive wave using Explicit, Alternating Direction Explicit (ADE), and Alternating Direction Explicit with Prediction Correction (ADEPC).
- Interpolation (if necessary): weighted inverse distance square and Theissen polygons.

7.3 MODEL SETTINGS

In the simulations presented here, the Green and Ampt with Redistribution (GAR) method is used for infiltration and subsurface analysis.

The GAR method approximates the solution of the Richard's equation for deep, well drained soil that is subject to multiple rainfall events (Downer and Ogden (2002)). The GAR approach is derived to overcome the weakness of the original Green and Ampt (Green and Ampt (1911)) model to simulate soil moisture redistribution and multiple rainfall events (Senarath et al. (2000)). The original Green and Ampt method provides a reasonable estimation of infiltration for the first rainfall event and underestimates the infiltration for subsequent rainfall events, as it does not consider redistribution of the soil moisture profile for the periods between the rainfall events (Senarath et al. (2000)). The GAR method, however, allows for long-term simulations which contain sequences of dry and wet periods. The soil moisture accounting scheme simulates the soil moisture redistribution along the soil profile during a runoff event, as well as the change in soil moisture due to evapotranspiration between rainfall events. It is noted that the GAR method assumes that during a particular rainfall event, evapotranspiration has a negligible effect on the soil moisture profile. By the end of the runoff event (between rainfall events), evapotranspiration dominates the changes in the soil moisture profile.

Evapotranspiration and overland flow analysis are performed using the Penman-Monteith (Monteith (1981)) and Alternating Direction Explicit (ADE) methods, respectively. Evapotranspiration reduces the basin water content including the surface storage and soil moisture in the unsaturated zone down to the specified root depth. The Penman-Monteith method, used in the model, is one of the most commonly used models for prediction of evapotranspiration from a vegetated land surface (Senarath et al. (2000)). Furthermore, in the developed model, the two-dimensional diffusive wave approximation of the St. Venant equations and 1-D explicit diffusive wave method are used for overland flow and channel routing.

The topographic information and hydrological properties of the watershed are represented by a 125 m^2 Cartesian grid. Note that Senarath et al. (2000) showed that this resolution is adequate to represent spatial properties of the watershed including topography, soil, hydrologic and land-use information. The 30 m^2 digital elevation model (DEM) of the watershed was obtained from the US Geological Survey (USGS) and used for model development. The dimensions of the channel cross-sections were compiled from historical surveys of the Agricultural Research Service (ARS). Overland roughness parameters were selected based on land-use information and literature sources at each grid pixel. Initial values of soil hydraulic conductivity, saturated hydraulic conductivity, effective porosity, soil suction head and evapotranspiration parameters including root depth and vegetation transmission coefficients are assigned based on Senarath et al. (2000) who developed a hydrologic model for the Goodwin Creek watershed used in this study. The selected initial values are then adjusted through model calibration.

The GSSHA model, used for simulations, is carefully calibrated as discussed in Habib et al. (2008) to ensure that the model will provide accurate representation of the rainfall-runoff processes in the watershed. The main calibration parameters included overland and channel roughness coefficients and soil infiltration parameters. The results show that the model was most sensitive to the soil saturated hydraulic conductivity and, to a lesser degree, to the overland and channel roughness coefficients (Habib et al. (2008)). For the details of the calibration procedure including the calibration periods, quantitative assessment of the model uncertainty and estimated error between the observed and simulated runoff, the reader is referred to Habib et al. (2008). It is noted that the final values of the calibrated model parameters were found to be within reasonable physical bounds which indicates that the model is likely to provide physically reasonable representation of the rainfall-runoff processes within the Goodwin Creek watershed.

7.4 APPLICATION

The rainfall ensembles generated for the Goodwin Creek watershed are used as input to the GSSHA hydrological model. For every rainfall event four different rainfall ensembles simulated using the Gaussian copula, t-copula, v-copula and random error model (E-Model) are used to estimate the output runoff of the Goodwin Creek watershed. For three rainfall events, Figures 7.1, 7.2 and 7.3 display the estimated runoff for the simulated ensembles (500 realizations) presented in Chapters 4, 5 and 6. Figures 7.1(a), 7.2(a) and 7.3(a) present the uncertainty of the output runoff (an ensemble of runoff estimates) obtained from simulated rainfall fields using the Gaussian copula. In Figures 7.1(b), 7.2(b) and 7.3(b) the simulated ensemble obtained from the t-copula model is used as input to the hydrologic model, whereas in Figures 7.1(c), 7.2(c) and 7.3(c) the v-copula ensemble is used for analysis. Finally, in Figures 7.1(d), 7.1(d) and 7.1(d) the input rainfall ensemble is generated using the random error model discussed, in Chapter 6. In the figures, the solid black and dashed red lines represent the runoff resulting from radar estimates and rain gauge measurements, respectively. The gray lines represent the estimated uncertainty of the output runoff (resulting from simulated rainfall realizations). As shown, the ensembles of runoff hydrographs encompass the hydrographs resulting from rain gauge measurements and radar estimates. In general, the output hydrographs resulting from gauge measurements and radar estimates can be significantly different. In such case, the estimated uncertainty of the output runoff can be useful for practical applications (e.g. risk assessment and decision making). Note that the aim of this chapter is solely to show an example application of stochastically simulated rainfall ensembles for hydrologic analysis. Other issues such as investigating the GSSHA model with respect to the input, uncertainty analysis of the hydrologic model and error propagation in hydrologic process are beyond the scope of this research, and thus, are not addressed here.

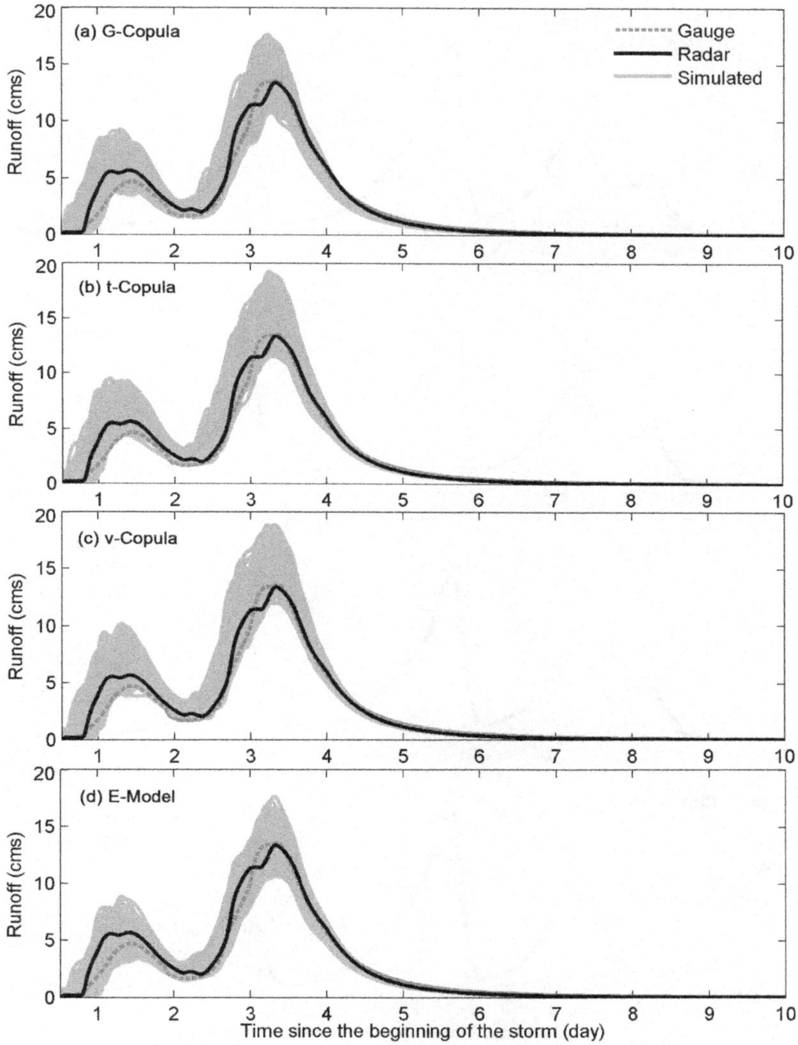

Figure 7.1 The resulting runoff of the rainfall ensembles obtained form the following models (GC2-II): (a) Gaussian copula (G-Copula); (b) t-copula; (c) v-copula; (d) random error model (E-Model).

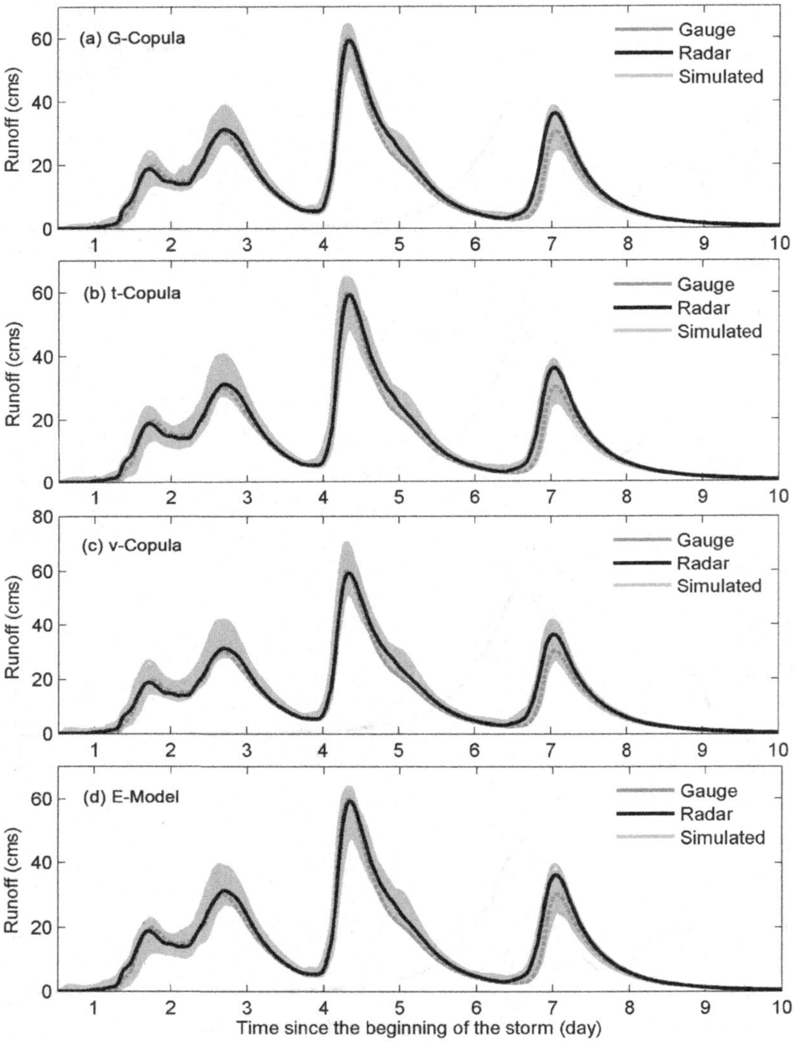

Figure 7.2 The resulting runoff of the rainfall ensembles obtained form the following models (GC4-II): (a) Gaussian copula (G-Copula); (b) t-copula; (c) v-copula; (d) random error model (E-Model).

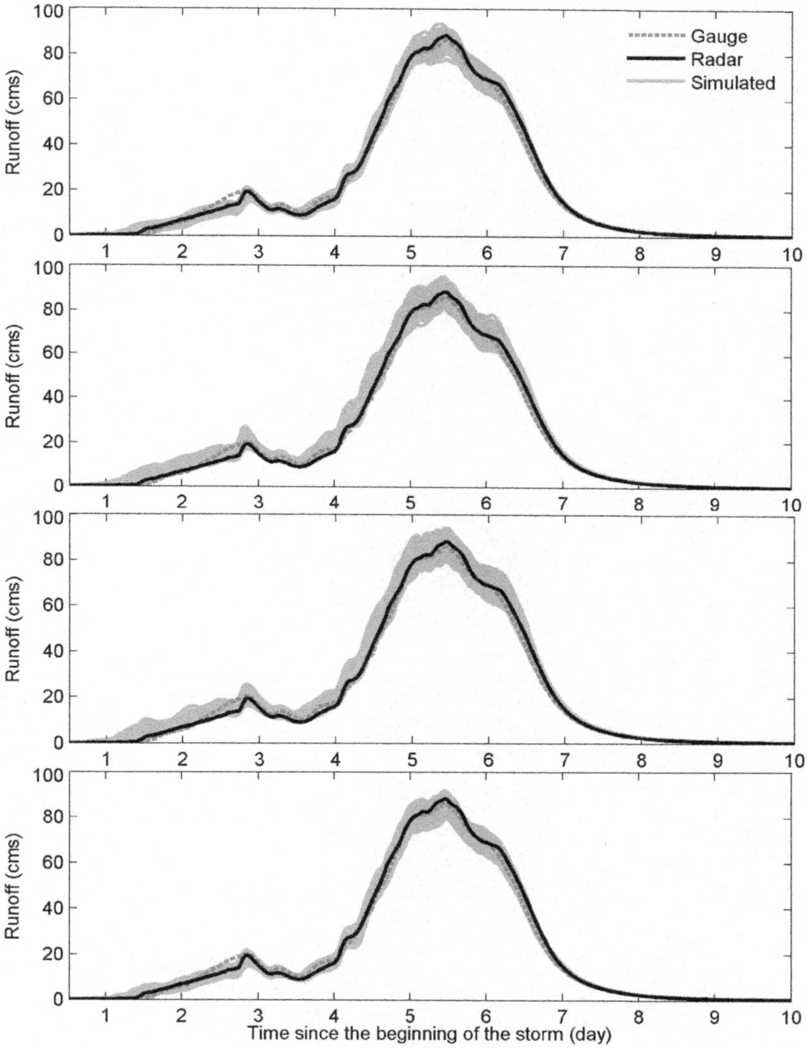

Figure 7.3 The resulting runoff of the rainfall ensembles obtained form the following models (GC5-II): (a) Gaussian copula (G-Copula); (b) t-copula; (c) v-copula; (d) random error model (E-Model).

CHAPTER 8

SUMMARY, DISCUSSION AND CONCLUSIONS

8.1 A BRIEF OVERVIEW

Recent technological advances in the field of remote sensing have led to an increase in available rainfall data on a regional and global scale. Specifically, the National Weather Service (NWS) Next Generation Weather Radar (NEXRAD) systems and NASA sponsored satellite missions provide the hydrometeorologic community with a myriad of new precipitation data. Such data have significant advantages over traditional sources of measurement. Unlike rain gauges, radar and satellite-based estimates are not limited by point sampling and can provide uninterrupted coverage regionally or globally with high spatial and temporal resolution. However, the advantages are limited due to the uncertainties associated with remotely sensed rainfall estimates (Hossain and Huffman (2008); Margulis et al. (2006); Dong et al. (2005)).

This study's purpose is to contribute to ongoing research regarding the uncertainty of remotely sensed rainfall estimates by using ensemble simulation. Rainfall ensembles can be used for various analyses. For example, with subsequent runs of a hydrological or meteorological model using an ensemble of rainfall realizations, one can investigate propagation of error in hydrological processes. Moreover, the uncertainty of a given hydrologic model resulting from uncertainties of the input precipitation can be assessed.

The aim of this study is to investigate and develop different stochastic techniques for simulation of rainfall ensembles through simulating random error fields and imposing them over remotely sensed rainfall estimates. Four different models are developed and discussed in this work. In the first and second models, two elliptical copulas (Gaussian and t-copula) are used to describe the dependence structure of radar rainfall error and to simulate multivariate rainfall error fields. In the third model, an asymmetrical v-transformed copula is employed for error simulations. This type of copula family is used to describe asymmetrical dependencies through the copula parameters. In the fourth model, rainfall fields are generated by perturbing rainfall estimates with two normally distributed error values. One is a purely random component, while the other component is proportional to the magnitude of the rainfall rates. In the first three models, having described the dependencies using copulas, the empirical distribution function of observed rainfall error is numerically approximated and applied to the simulated error fields. In terms of their distribution functions, the resultant simulated realizations are similar to those of the observed. In the fourth model, however, the error is assumed to be normally distributed. In all the models, available observations of radar rainfall error (the differences between radar estimates and rain gauge measurements) are used to condition the simulated fields on observations.

In order to examine reliability and performance of the presented models, several case studies are presented for the Little Washita and Goodwin Creek Watersheds. Both radar reflectivity data (Stage II) as well as Stage IV Next Generation Weather Radar (NEXRAD) multi-sensor precipitation estimates are used as input to the models. Regarding the

MPE data, at present, different procedures have been introduced and implemented by the National Weather Service to improve the quality of the multi-sensor precipitation estimates. These methods include merging of radar and gage estimates, mean-field bias adjustments and local bias adjustments. However, the inherent limitations of radar-based estimates to accurately measure surface rainfall makes it necessary to assess the underlying uncertainties of the MPE data and account for them. Therefore, in addition to the radar reflectivity estimates, MPE data are also used for simulations.

8.2 DISCUSSION AND CONCLUSIONS

In previous chapters, the simulated rainfall fields obtained from different models are compared with original radar estimates with respect to statistical properties, extreme values and spatio-temporal dependencies. In the following, the presented models are compared in more detail with respect to their statistical properties. As mentioned before, cross validation (Picard and Cook (1997); Efron and Tibshirani (1990)) is used to test how well the copula fits the available observations. Figure 8.1 displays the results of cross validation (MAE estimator) for the models when: (a) all gauges; (b) 50% of the gauges; (c) 12% LW (25% GC) rain gauges are used in the analysis. One can see that in Figures 8.1(a) and 8.1(b), the differences between the MAE values are not considerable. However, as the number of rain gauges reduces, the differences become significant. In Figure 8.1(c), where 12% LW (25% GC) gauges are used, the MAE values for the t-copula model are less than the Gaussian and v-copula models. Furthermore, the MAE for the v-copula model is found to be less than the Gaussian copula particularly in Figure 8.1(c). In terms of the copula fit, the figures indicate that the t-copula family fits better to the observations; however, this case may not be generalized, as the choice of copula family is known to be problem dependent.

The presented models are also investigated with regard to preserving the spatial dependence structure. Figure 8.2 shows the Mean Absolute Error (MAE) of the correlation matrices of the simulated rainfall fields

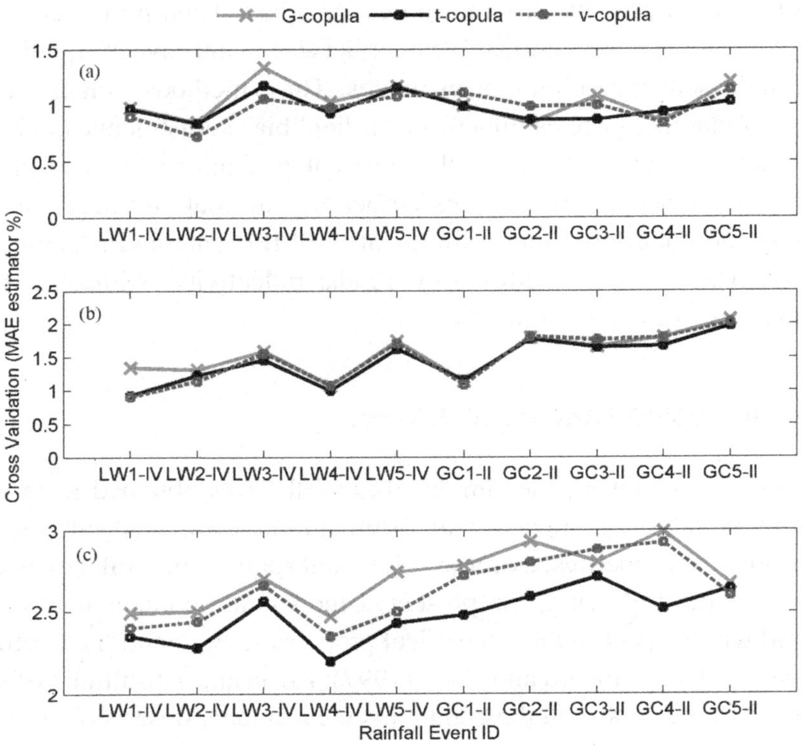

Figure 8.1 The results of cross validation (MAE estimator, %) for the models when: (a) all gauges; (b) 50% of the gauges; (c) 12% of LW (25% of GC) gauges; were used in the analysis.

(500 realizations) with respect to the correlation matrix of the observed radar rainfall data. Figure 8.2(a) presents the MAE values when all gauges are included in the simulations. Figures 8.2(b) and Figure 8.2(c) display the MAE values when 50% of the gauges and 12% of LW (25% of GC) rain gauges are used for the analysis. In the figure, the black solid, blue dashed and x-marked lines represent t-copula, v-copula and Gaussian copula models, respectively. As shown, the MAE values are similar for all the copula-based models. By reducing the number of rain gauges, the MAE values increase (as shown in Figures 8.2(b) and 8.2(c)), while the MAE values of all the models remain similar.

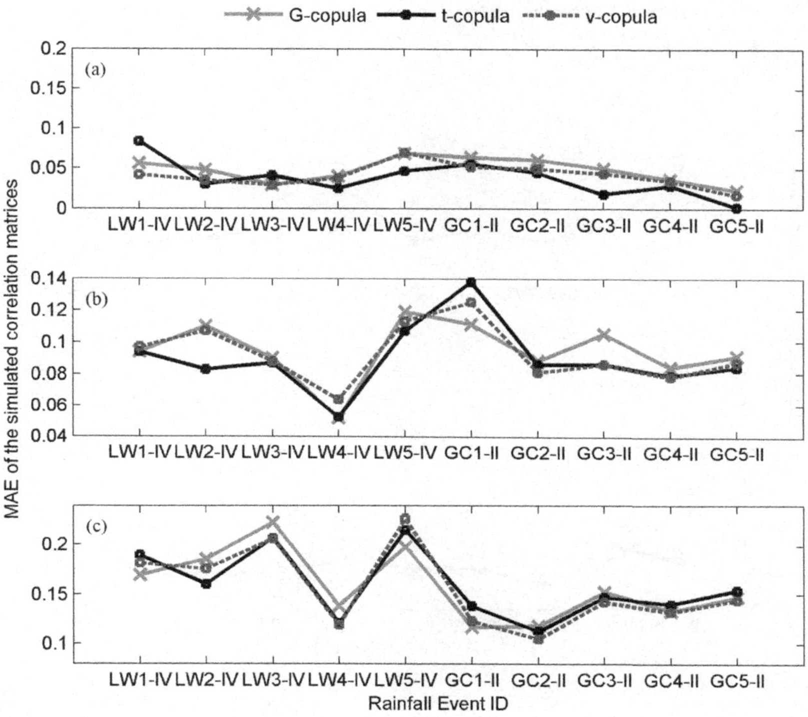

Figure 8.2 MAE of the simulated correlation matrices.

This study addresses the capability of the presented models to simulate the number of extreme occurrences above the 90th percentile and 75th percentile thresholds. Figure 8.3(a) plots the error (%) in the number of extreme occurrences above the 90th percentile threshold. As shown, the t-copula (dashed blue line) and v-copula (solid black line) models exhibit less error in the number of simultaneous extremes. The figure also indicates that the t-copula model shows slightly lower error. For the lower threshold (75th percentile), Figure 8.3(b) displays the error (%) in the simulated extreme occurrences. In this figure, the differences are less significant than those shown in Figure 8.3(a). Nevertheless, Figure 8.3(b) shows that the t-copula and v-copula models perform better than both the Gaussian copula model and random error

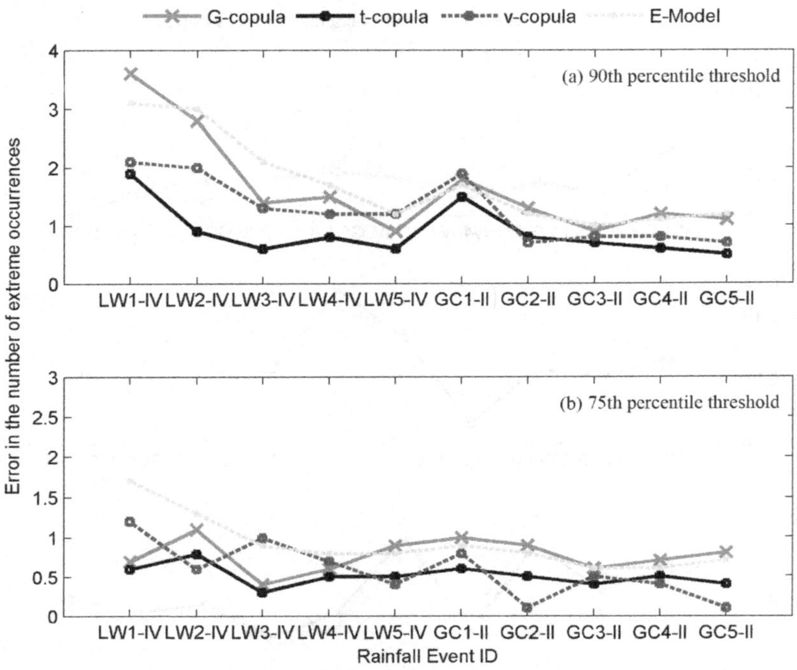

Figure 8.3 Error (%) in the number of extreme occurrences for the models: (a) 90th percentile threshold; (b) 75th percentile threshold.

model. Therefore, when simultaneous extremes are considered, the t-copula or v-copula family is preferred over the Gaussian copula.

In this study, the models are also validated by using different numbers of gauges for simulations to investigate whether the estimated uncertainty enclose rain gauge measurements when less numbers of gauges were available. This issue was considered by counting the number of time steps (n_{out}) where the ground reference measurements did not fall within the estimated uncertainty. For the presented models, Figure 8.4 displays the n_{out} values when: 8.4(a) all gauges; 8.4(b) 50% of the gauges; 8.4(c) 12% of LW (25% of GC) gauges are used for simulations. As shown in the figures, the n_{out} values for the random error model (E-Model) are less than the other models. This could be because of the fact that in the random error model, the CDF of the observed error is not applied on the simulated values which results in wider uncertainty bound. It is pointed out that as the number of gauges decreases, the sample size of the observed error decreases. This may, but not necessarily, result in a shorter range of error (the difference between minimum and maximum error). One can intuitively conclude that any change in the error range, and thus the empirical distribution of error, will affect the estimated uncertainty. The figures indicate that reducing the number of gauges results in an increase in the value of n_{out}. Nonetheless, with a few rain gauges available, the simulated ensembles obtained from the presented models, enclose approximately 93% to 96% of the ground reference measurements. It is also noted that the values of n_{out} for the t-copula and v-copula models are similar, and both are slightly better than the Gaussian copula model.

Verification of the 95% confidence intervals for the number of time steps that the estimated uncertainty did not enclose the rain gauge measurements is also provided. Figure 8.5(a) plots the n_{out} (%) values when all gauges are included. Figures 8.5(b) and (c) show similar graphs when 50% of the gauges and 12% of LW (25% of GC) gauges are used for simulations, respectively. The figure confirms that the E-Model results in the least error in the n_{out} values. The differences become significant when few gauges are included. This implies that for poorly gauged watersheds, the E-Model may be superior over the others.

For practical applications, the duration of the simulation could be an important factor, particularly, if simulation in real-time is required (e.g. for analysis within the next couple of hours). Figure 8.6 displays the CPU time (seconds) for the presented models. One can see that the v-copula model is computationally more demanding, which results from parameter estimation. The figure also indicates that the Gaussian copula and random error models are relatively faster than the rest, and that the t-copula model is not significantly slower than the Gaussian copula model. It is noted that the figure shows the CPU time as a reference for comparison and the actual time required for the analysis is significantly longer.

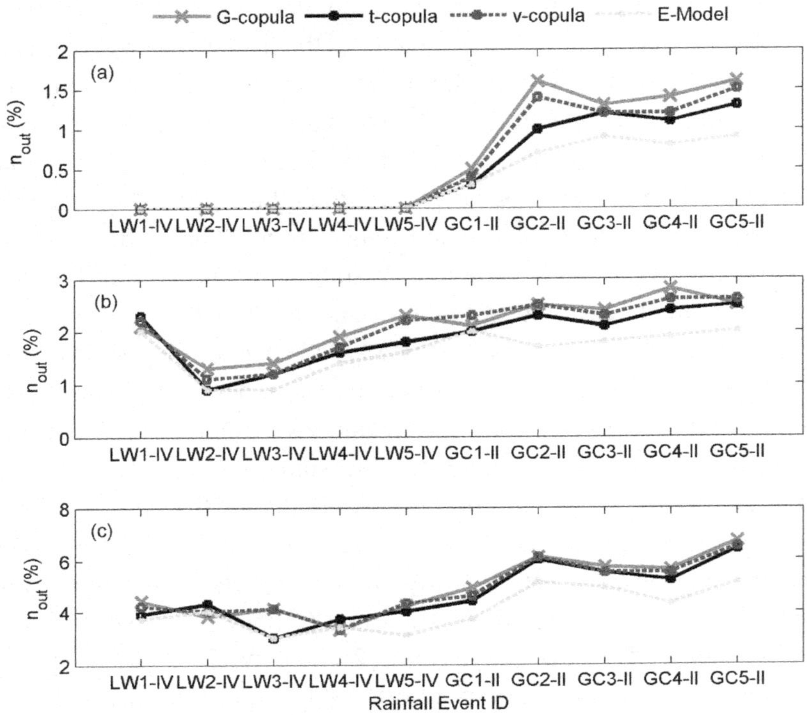

Figure 8.4 The number of time steps, n_{out} (%), that the estimated uncertainty did not encompass the rain gauge measurements: (a) all gauges; (b) 50% of the gauges; (c) 12% of LW (25% of GC) gauges.

The results indicate that the presented models are quite similar with respect to the dependence structure. However, using the t-copula model may have significant advantages over the other models, particularly, with respect to extremes. Furthermore, the cross validation analysis showed that the t-copula family fits the observations better than the other models. Nevertheless, it is often more advantageous to substitute a complex copula that fits the data better with a simple and practical copula. Provided that the effects of this substitution can be quantified, one may opt to use a simple copula that provides a sufficiently good statistical approximation of the actual copula (e.g. Gaussian copula) rather than a more sophisticated copula family (e.g. t-copula). In

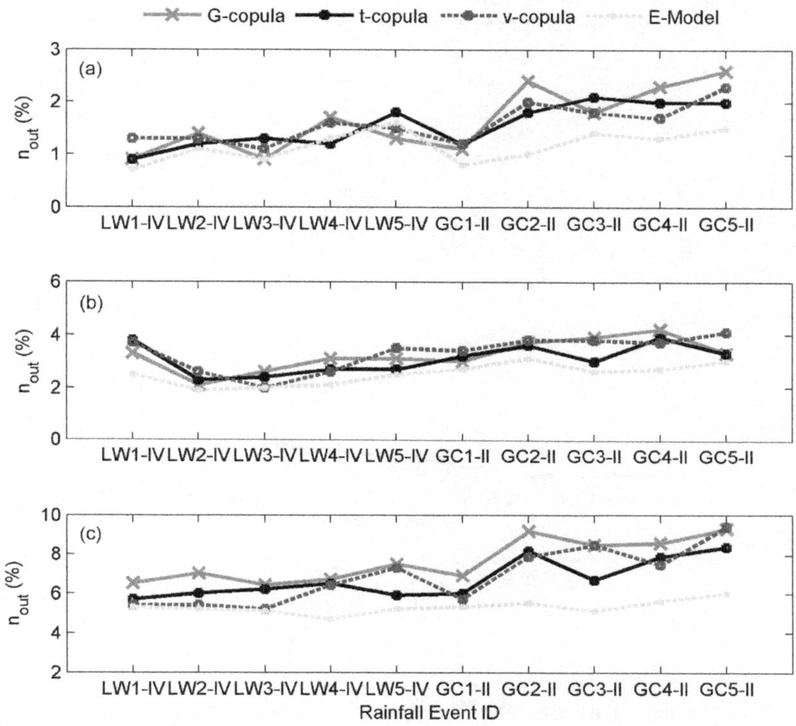

Figure 8.5 Verification of the 95% confidence intervals for the number of time steps, n_{out} (%), that the estimated uncertainty did not encompass the rain gauge measurements: (a) all gauges; (b) 50% of the gauges; (c) 12% of LW (25% of GC) gauges.

principle, the choice of copula family is based on the empirical evidence as well as the technical constraints, such as the number of parameters essential to describe the copula. Moreover, the issue of choosing a copula family may be considered problem dependent, and therefore cannot be simply generalized. The analyses showed that with respect to the simulation time and the width of the estimated uncertainty the random error model is the most practical. Overall, the t-copula and random error models seem to be promising for future in-depth research. Nonetheless, all the presented models are able to simulate rainfall ensembles with similar statistical characteristics to those of the ground reference measurements.

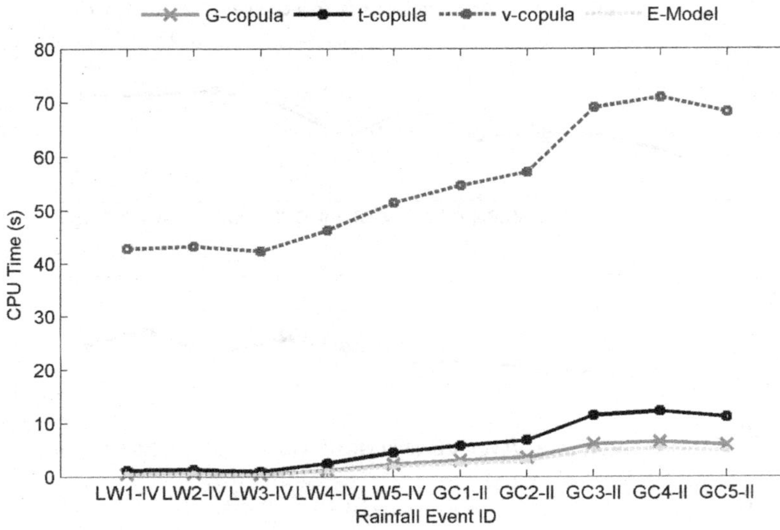

Figure 8.6 CPU Time (s).

The presented models are subject to various limitations. For example, in the random error model the error components are assumed to be normally distributed. Such an assumption has been tested and used by Ciach et al. (2007) and Villarini et al. (2009). The analyses presented here are performed on rainfall data with high temporal resolution (15 min), which may have some drawbacks (e.g. random errors in tipping-bucket measurements). Additionally, aggregation of data in time may result in the loss of temporal variability. Further research including simulations over different spatial and temporal scales and different gauge network setups are required to verify the appropriateness and applicability of this assumption. Another limitation concerning the implementation of this method is that unreliable surface gauge measurements may result in erroneous parameter estimation, and thus create an unrealistic ensemble and inaccurate measures of uncertainty. Also, it is worth pointing out that an individual gauge in a radar pixel is used to represent the true areal average rainfall over a pixel size, which may not be very accurate. However, based on the available data and the spatial resolution of radar data, this is the best possible approximation of the true areal average rainfall values. Based on previous research which shows that the temporal autocorrelations of the radar rainfall error are insignificant (Habib et al. (2008)), no explicit accounting for error autocorrelation is included in the models. That is, the term ϵ in the Gaussian copula, t-copula and v-copula models and both error terms, ϵ_1 and ϵ_2, in the random error model are assumed to have no temporal self-correlation. However, the results show that the self-correlation of the radar field will be carried forward when simulated fields are imposed over radar rainfall estimates.

It is hoped that the developed models and the results of this research can be used to assess the uncertainties associated with radar-based rainfall estimates, as it is believed that with accurate information about surface rainfall and its associated uncertainties, hydrologists and meteorologists have the potential to improve hydrologic predictions and global climate studies. Such models may also provide the required tools to investigate error propagation in modeling hydrologic processes and give the ability to analyze the uncertainty of hydrologic models resulting from input rainfall uncertainty. Furthermore, using ensemble analysis instead of a single realization, one can evaluate flood prediction uncertainty and its associated risks for a given precipitation by using an ensemble of precipitation estimates. The models introduced in this work can also be used for simulation of rainfall ensembles for climate change studies such as investigating long term changes in the hydrologic cycle of a specific watershed.

8.3 PERSPECTIVE OF FUTURE WORK

Based on the analysis carried out in this study, the results discussed above, and the scope of this research, the following recommendations for future research are suggested:

- In recent years, in addition to the NEXRAD radar data, several NASA sponsored satellite missions such as the Earth Observing System (EOS), Aqua satellite, Tropical Rainfall Measuring Mission (TRMM) and the anticipated Global Precipitation Mission

(GPM) have and will provide valuable hydrologic data. The presented models can be tested and verified using satellite data that are also subject to the uncertainties and limitations associated with radar rainfall estimates.

- Theoretically, there are an infinite number of possible copulas families (Salvadori et al. (2007); Nelsen (2006); Genest and MacKay (1986); Genest and Favre (2007)) that can be used for describing the dependencies. However, no extensive empirical study has determined the families of copulas that are appropriate for weather variables. Future research in this area can provide valuable information for practical applications.

- The problem of accounting for spatial and temporal dependencies of radar error warrants further investigations. As discussed before, no temporal autocorrelation was assumed for the simulated error fields. In future research, one may consider including the temporal autocorrelation of error in the model using different methods such as combining the Markov approach with a copula-based model.

- An estimation of the variability of model parameters with respect to different types of storms (e.g. stratiform, convective, etc.). Such analyses, however, require extensive high quality radar rainfall estimates and rain gauge measurements in order to obtain representative information on radar rainfall error.

Bibliography

Abdous, B., Ghoudi, K., Rémillard, B., 2003. Nonparametric weighted symmetry test. Can. J. Stat. 31 (4), 357381.

AghaKouchak, A., Bárdossy, A., Habib, H., 2010a. Copula-based uncertainty modeling: Application to multi-sensor precipitation estimates. Hydrological Processes 24, doi: 10.1002/hyp.7632.

AghaKouchak, A., Habib, E., Bárdossy, A., 2010b. Modeling radar rainfall estimation uncertainties: random error model. Journal of Hydrologic Engineering 15 (4), doi:10.1061/(ASCE)HE.1943-5584.0000185.

Alonso, C., Binger, R., 2000. Goodwin creek experimental watershed: A unique field laboratory. J. Hydraul. Engng ASCE 126, 174–177.

Anagnostou, E., Krajewski, W., 1997. Simulation of radar reflectivity fields: Algorithm formulation and evaluation. Water Resources Research 33, 14191428.

Arnaud, P., Bouvier, C., Cisner, L., Dominguez, R., 2002. Influence of rainfall spatial variability on flood prediction. Journal of Hydrology 260, 216230.

Austin, P., 1987. Relation between measured radar reflectivity and surface rainfall. Mon. Wea. Rev. 115, 1053–1070.

Bárdossy, A., 2006. Copula-based geostatistical models for groundwater quality parameters. Water Resources Research 42 (11), W11416.1–W11416.12, doi:10.1029/2005WR004754.

Bárdossy, A., Li, J., 2008. Geostatistical interpolation using copulas. Water Resources Research 44 (7), W07412.1–W07412.15, doi:10.1029/2007WR006115.

Bardossy, A., Plate, E., 1992. Spacetime models for daily rainfall using atmospheric circulation patterns. Water Resources Research 28, 12471259.

Bastin, L., Fisher, P., Wood, J., 2002. Visualizing uncertainty in multispectral remotely sensed imagery. Computers & Geosciences 28 (3), 337–350.

Battan, L., 1973. Radar Observation of the Atmosphere. The University of Chicago Press.

Bell, V., Moore, R., 2000. The sensitivity of catchment runoff models to rainfall data at different spatial scales. Hydrology and Earth System Sciences 4, 653–667.

Bernardo, J., Smith, A., 2000. Bayesian Theory. John Wiley and Sons, New York.

Bouye, E., Durrleman, V., Nikeghbali, A., Riboulet, G., Roncalli, T., 2000. Copulas for finance: A reading guide and some applications. Working paper Groupe de recherche Opérationelle Credit Lyonnais.

Brandt, S., 1999. Data Analysis, Statistical and Computational Methods for Scientists and Engineers. Third Edition, Springer.

Bras, R., Rodriguez-Iturbe, I., 1976. Rainfall generation: a nonstationary time varying multidimensional model. Water Resources Research 12, 450456.

Buishand, T., Brandsma, T., 2001. Multi-site simulation of daily precipitation and temperature in the rhine basin by nearest-neighbor resampling. Water Resources Research 37, 2761–2776.

Carpenter, T., Georgakakos, K., 2004. Impacts of parametric and radar rainfall uncertainty on the ensemble streamflow simulation. Journal of Hydrology 298, 242–266.

Chandler, R., Wheater, H., 2002. Analysis of rainfall variability using generalized linear models: A case study from the west of ireland. Water Resources Research 38 (10), doi:10.1029/2001WR00090.

Chow, V., Maidment, D., Mays, L., 1988. Applied Hydrology. McGraw-Hill Science Engineering.

Ciach, G., Habib, E., Krajewski, W., 2003. Zero-covariance hypothesis in the error variance separation method of radar rainfall verification. Advances in Water Resources 26, 573–580.

Ciach, G., Krajewski, W., 1999. On the estimation of radar rainfall error variance. Advances in Water Resources 22, 585595.

Ciach, G., Krajewski, W., Villarini, 2007. Product-error-driven uncertainty model for probabilistic quantitative precipitation estimation with nexrad data. Journal of Hydromet. 8, 1325–1347.

Clark, M., Gangopadhyay, S., Brandon, D., Werner, K., Hay, L., Rajagopalan, B., Yates, D., 2004a. A resampling procedure for generating conditioned daily weather sequences. Water Resources Research 40 (4), W04304.1–W04304.15, doi:10.1029/2003WR002747.

Clark, M., Gangopadhyay, S., Hay, L., Rajagopalan, B., Wilby, R., 2004b. Schaake shuffle: A method for econstructing space-time variability in forecasted precipitation and temperature fields. Journal of Hydrometeorology 5, 243 262.

Corradini, C., Singh, V., 1985. Effect of spatial variability of effective rainfall on direct runoff by geomorphoiogic approach. Journal of Hydrology 81, 27–43.

Cox, D., Hinkley, D., 1974. Theoretical Statistics. Chapman & Hall.

Crosetto, M., Ruiz, J., Crippa, B., 2001. Uncertainty propagation in models driven by remotely sensed data. Remote Sensing of Environment 76 (3), 373–385.

Crum, T., Alberty, R., 1993. The wsr-88d and the wsr-88d operational support facility. Bull. Amer. Meteor. Soc. 74, 1669–1687.

DallÁglio, G., 1991. Fréchet classes: The beginnings. In: Advance in Probability Distributions with Given Marginals (G. DallÁglio, S. Kotz, and G. Salinetti, Eds.), Kluwer Academic, Dordrecht.

Davidson, R., MascKinnon, J., 1993. Estimation and inference in econometrics. Oxford University Press, Oxford.

Dawdy, D., Bergman, J., 1969. Effect of rainfall variability on strearn-flow simulation. Water Resources Research 5, 958–966.

De Michele, C., Salvadori, J., 2003. A generalized pareto intensity-duration model of storm rainfall exploiting 2-copulas. J. Geophy. Res. Atmos. 108 (D2), 111, doi:10.1029/2002JD002534.

Deardorff, J., 1997. Parameterization of ground-surface moisture-content for use in atmospheric prediction models. Journal of Applied Meteorology 16 (11), 1182–1185.

Dixon, M., Kessinger, C., Hubbert, J., 2005. Echo classification and spectral processing for the discrimination of clutter from weather. Preprints, 32nd Conference on Radar Meteorology, American Meteorological Society, Albuquerque, NM, 24-29 September 2005.

Doelling, I., Joss, J., Riedl, J., 1998. Systematic variations of zr relationships from drop size distributions measured in northern germany during seven years. Atmos. Res. 4748, 635649.

Dong, J., Walker, J., Houser, P., 2005. Factors affecting remotely sensed snow water equivalent uncertainty. Remote Sensing of Environment 97 (1), 68–82.

Downer, C., Ogden, F., 2002. Gssha user's manual, gridded surface subsurface hydrologic analysis version 1.43 for wms 6.1. Tech. rep., Engineer Research and Development Center Technical Report, Vicksburg, MS.

Dupuis, D., 2007. Using copulas in hydrology: Benefits, cautions, and issues. Journal of Hydrologic Engineering 12 (4), 381393.

Efron, B., Tibshirani, R., 1990. Improvements on cross-validation: The .632 + bootstrap method. Journal of the American Statistical Association 92 (438), 548–560.

Elliott, R., Schiebe, F., Crawford, K., Peter, K., Puckett, W., 1993. A Unique Data Capability for Natural Resources Studies. International Winter Meeting, American Society of Agricultural Engineers, Chicago, IL, Dec. 14-17, 1993, paper No. 932529.

Embrechts, P., Lindskog, F., McNeil, A., 2001. Modelling Dependence with Copulas and Applications to Risk Management. Department of Mathematics, ETHZ, Zurich, Switzerland.

Fang, H.-B., Fang, K.-T., Kotz, S., 2002. The meta-elliptica distribution with given marginals. Journal of Multivariate Analysis 82, 116.

Faures, J.-M., Goodrich, D., Woolhiser, D., S., S., 1995. Impact of small-scale spatial rainfall variability on runoff modeling. J. Hydrology 173, 309–326.

Favre, A.-C., El Adlouni, S., Perreault, L., Thiémonge, N., Bobée, B., 2004. Multivariate hydrological frequency analysis using copulas. Water Resources Research 40, 112.

Fowler, H., Kilsby, C., O'Connell, P., Burton, A., 2005. A weather-type conditioned multi-site stochastic rainfall model for the generation of scenarios of climatic variability and change. Journal of Hydrology 308, 50–66.

Genest, C., Favre, A., Béliveau, J., Jacques, C., 2007. Metaelliptical copulas and their use in frequency analysis of multivariate hydrological data. Water Resources Research 43 (9), W09401.1–W09401.12, doi:10.1029/2006WR005275.

Genest, C., Favre, A.-C., 2007. Everything you always wanted to know about copula modeling but were afraid to ask. Journal of Hydrologic Engineering 12 (4), 347–368.

Genest, C., MacKay, R., 1986. Archimedean copulas and families of bidimensional laws for which the marginals are given. Canadian Journal of Statistics 14, 145–159.

Germann, U., Berenguer, M., Sempere-Torres, D., 2009. Ensemble radar precipitation estimation for hydrology in a mountainous region. Quarterly Journal of the Royal Meteorological Society 135 (639), 445–456.

Germann, U., Berenguer, M., Sempere-Torres, D., Salvade, G., 2006. Ensemble radar precipitation - a new topic on the radar horizon. In: Proceeding of the 4th European Conference on Radar in Meteorology and Hydrology ERAD. Barcelona, Spain, 18-22 September 2006, pp. 559–562.

Gomez-Hernandez, J., Wen, X., 1998. To be or not to be multi-gaussian? a reflection on stochastic hydrogeology. Advances in Water Resources 21, 47–61.

Goodrich, D., Faures, J., Woolhiser, D., Lane, L., Sorooshian, S., 1995. Measurement and analysis of small-scale convective storm rainfall variability. Journal of Hydrology 173, 283308.

Green, W., Ampt, G., 1911. Studies of soil physics, part i the flow of air and water through soils. J. Ag. Sci. 4, 1–24.

Haberlandt, U., Ebner von Eschenbach, A.-D., Buchwald, I., 2008. A space-time hybrid hourly rainfall model for derived flood frequency analysis. Hydrology and Earth System Sciences 12, 1353–1367.

Habib, E., Aduvala, A., Meselhe, E., 2008. Analysis of radar rainfall error characteristics and implications for streamflow simulations uncertainty. J. Hydrologic Sciences 53 (3), 568–587.

Habib, E., Ciach, G., Krajewski, W., 2004. A method for filtering out raingauge representativeness errors from the verification distributions of radar and raingauge rainfall. Journal of Advances in Water Resources 27 (10), 967–980.

Habib, E., Henschke, A., Adler, R., 2009. Evaluation of tmpa satellite-based research and real-time rainfall estimates during six tropical-related heavy rainfall events over louisiana, usa. Atmospheric Research 94 (3), 373–388, doi:10.1016/j.atmosres.2009.06.015.

Haddad, J., Nimah, M., Farajallah, N., 2007. Modeling annual rainfall: a robust maximum likelihood approach. Environmetrics 18 (1), 101–105.

Hamlin, M., 1983. The significance of rainfall in the study of hydrological processes at basin scale. Journal of Hydrology 65, 73–94.

Hollander, M., Wolfe, D., 1973. Nonparametric Statistical Methods. John Wiley and Sons, New York.

Hong, Y., Hsu, K., Moradkhani, H., Sorooshian, S., 2006. Uncertainty quantification of satellite precipitation estimation and monte carlo assessment of the error propagation into hydrologic response. Water Resources Research 42 (8), w08421, doi:10.1029/2005WR004398.

Hossain, F., Anagnostou, E., 2005. Numerical investigation of the impact of uncertainties in satellite rainfall estimation and land surface model parameters on simulation of soil moisture. Advances in Water Resources 28 (12), 1336–1350.

Hossain, F., Huffman, G., 2008. Investigating error metrics for satellite rainfall data at hydrologically relevant scales. Journal of Hydrometeorology 9 (3), 563–575.

Hunter, S., 1996. Wsr-88d radar rainfall estimation: capabilities, limitations and potential improvements. NWA Digest 20, 26–36.

Hutchinson, M., 1995. Stochastic space time weather models from ground based data. Agricultural and Forest Meteorology 73, 237264.

IPCC, 2007. Climate Change 2007: Impacts, Adaptation, and Vulnerability. Exit EPA Disclaimer Contribution of Working Group II to the Third Assessment Report of the Intergovernmental Panel on Climate Change [Parry, Martin L., Canziani, Osvaldo F., Palutikof, Jean P., van der Linden, Paul J., and Hanson, Clair E. (eds.)]. Cambridge University Press, Cambridge, United Kingdom.

Joe, H., 1997. Multivariate Models and Dependence Concepts. Chapman Hall, London.

Joe, H., Xu, J., 1996. The estimation method of inference functions for margins for multivariate models. Technical Report 166, Department of Statistics, University of British Columbia, Vancouver, Canada.

Jordan, P., Seed, A., Weinmann, P., 2003. A stochastic model of radar measurement errors in rainfall accumulations at catchment scale. J. Hydrometeorol. 4 (5), 841–855.

Journel, A., Alabert, F., 1989. Non-gaussian data expansion in the earth sciences. Terra Nova 1, 123–134.

Journel, A., Huijbregts, C., 1978. Mining geostatistics. Academic Press, New York.

Kelly, k., Krzysoztofowicz, R., 1997. A bivariate meta-gaussian density for use in hydrology. Stochastic Environ. Res. Risk Assess 11 (1), 1731.

Khalilia, M., Leconte, R., 2007. Efficient stochastic generation of multi-site synthetic precipitation data. Journal of Hydrology 345, 121–133.

Kim, T., Ahn, H., Chung, G., Yoo, C., 2008. Stochastic multi-site generation of daily rainfall occurrence in south florida. Stochastic Environmental Research and Risk Assessment 22 (6), 705717, doi: 10.1007/s00477-007-0180-8.

Kitchen, M., Blackall, R., 1992. Representativeness errors in comparisons between radar and gauge measurements of rainfall. Journal of Hydrology 134, 1333.

Kotz, S., 1997. Some remarks on copulas in relation to modern multivariate analysis. International Symposium on Contemporary Multivariate Analysis and Its Applications, Hong Kong, 1997.

Krajewski, W., Georgakakos, K., 1985. Synthesis of radar rainfall data. Water Resources Research 21, 764–768.

Kuhn, G., Khan, S., Ganguly, A., 2007. Geospatial-temporal dependence among weekly precipitation extremes with applications to observations and climate model simulations in south america. Journal of Advances in Water Resources 30, 2401–2423.

Lall, U., Rajagopalan, B., Tarboton, D., 1996. A nonparametric wet/dry spell model for resampling daily precipitation. Water Resources Research 32, 28032823.

Lovejoy, S., Schertzer, D., 1990. Fractals, raindrops and resolution dependence of rain measurements. Journal Applied Meteorology 29, 11671170.

Lucieer, A., Kraak, M., 2004. Interactive and visual fuzzy classification of remotely sensed imagery for exploration of uncertainty. International Journal of Geographical Information Science 18 (5), 491–512.

Malevergne, Y., Sornette, D., 2003. Testing the gaussian copula hypothesis for financial assets dependences. In: Quantitative Finance Volume 3, 231250.

Margulis, S., Entekhabi, D., McLaughlin, D., 2006. Spatiotemporal disaggregation of remotely sensed precipitation for ensemble hydrologic modeling and data assimilation. Journal of Hydrometeorology 7 (3), 511–533.

McLeish, D., Small, C., 1988. The Theory and Applications of Statistical Inference Functions. vol. 44 of Lecture Notes in Statistics, Springer-Verlag, New York.

Mehrotra, R., Srikanthan, R., Sharma, A., 2006. A comparison of three stochastic multi-site precipitation occurrence generators. Journal of Hydrology 331, 280292.

Melchiori, M., 2003. Which Archimedean Copula is the right one? Third Version, YeldCurve e-Journal, September 2003.

Miller, C., Edwards, P., 2001. Changing the Atmosphere: Expert Knowledge and Environmental Governance. MIT Press.

Monteith, J., 1981. Evaporation and surface temperature. Q. J. R. Meteorol. Soc. 107, 1–27.

Montopoli, M., Marzano, F., 2007. Maximum-likelihood retrieval of modeled convective rainfall patterns from midlatitude c-band weather radar data. IEEE Transactions on GEOSCIENCE and Remote Sensing 45 (7), 2403–2416.

Nelsen, R., 2006. An Introduction to Copulas (Springer Series in Statistics). Springer Verlag, New York.

Obled, C., Wendling, J., Beven, K., 1994. The sensitivity of hydrological models to spatial rainfall patterns: an evaluation using observed data. Journal of Hydrology 159, 305–333.

Ogden, F., Julien, P., 2002. CASC2D: A Two-Dimensional, Physically-Based, Hortonian Hydrologic Model. In: Mathematical Models of Small Watershed Hydrology and Applications, (Singh, V.P. and Frevert, D., eds.), Water Resources Publications, Littleton, Colorado, USA.

Ogden, F., Saghafian, B., 1997. Green and ampt infiltration and redistribution. J. Irrigation and Drainage Eng. 123, 386–393.

Pegram, G., Clothier, A., 2001. Downscaling rainfields in space and time using the string of beads model in time series mode. Hydrology and Earth System Sciences 5 (2), 175–186.

Petersen-Ø verleir, A., 2004. Accounting for heteroscedasticity in rating curve estimates. Journal of Hydrology 292, 173181.

Picard, R., Cook, D., 1997. Cross-validation of regression models. Journal of the American Statistical Association 79 (387), 575–583.

Purwono, Y., 2005. Copula inference for multiple lives analysis - preliminaries. International Actuarial Association: 13th EAA Conference, 12-15 September 2005, Bali, Indonesia.

Rajagopalan, B., Lall, U., 1999. A k-nearest neighbour simulator for daily precipitation and other weather variables. Water Resources Research 35, 3089–3101.

Renard, B., Lang, M., 2007. Use of a gaussian copula for multivariate extreme value analysis: Some case studies in hydrology. Advances in Water Resources 30, 897–912.

Richards, L., 1931. Capillary conduction of liquids through porous mediums. Physics 1 (5), 318–333, doi: 10.1063/1.1745010.

Rico-Ramirez, M., Cluckie, I., Shepherd, G., Pallot, A., 2007. A high resolution radar experiment on the island of jersey. Meteorol. Applications 14, 117–129.

Rodda, J., 1967. The systematic errors in rainfall measurement. J. Inst. Water Eng. 21, 173–177.

Ruddiman, W., 2005. Plows, Plagues and Petroleum: How Humans Took Control of Climate. Princeton University Press.

Salvadori, G., DeMichele, C., Kottegoda, N., Rosso, R., 2007. Extremes in Nature: An approach using copulas. Springers, Berlin.

Schölzel, C., Friederichs, P., 2008. Multivariate non-normally distributed random variables in climate research introduction to the copula approach. Nonlinear Processes in Geophysics 15 (5), 761–772.

Schuurmans, J., Bierkens, M., 2007. Effect of spatial distribution of daily rainfall on interior catchment response of a distributed hydrological model. Hydrology and Earth System Sciences 11, 677–693.

Schweizer, B., 1991. Thirty years of copulas. In: Advance in Probability Distributions with Given Marginals (G. DallÁglio, S. Kotz, and G. Salinetti, Eds.), Kluwer Academic, Dordrecht.

Seber, G., Wild, C., 1998. Nonlinear Regression. John Wiley and Sons, Chichester.

Seed, A., Srikanthan, R., 1999. A space and time model for design storm rainfall. J. Geophysical Research 104 (D24), 3162331630.

Seliga, T., Aron, G., Aydin, K., White, E., 1992. Simulation using radar rainfall rates and a unit hydrograph model (syn-hyd) applied to greve watershed. American Meteorological Society, 25th International Conference on Radar Hydrology, Paris, France, 587–590.

Senarath, S., Ogden, F., Downer, C., Sharif, H., 2000. On the calibration and verification of distributed, physically-based, continouous, hortonian hydrologic models. Water Resources Research 36 (6), 1495–1510.

Serinaldi, F., 2008. Analysis of inter-gauge dependence by kendall's τ, upper tail dependence coefficient, and 2-copulas with application to rainfall fields. Stochastic Environmental Research and Risk Assessment 22, 671688.

Serinaldi, F., 2009. Copula-based mixed models for bivariate rainfall data: An empirical study in regression perspective. Stochastic Environmental Research and Risk Assessment 23 (5), 677–693, doi: 10.1007/s00477-008-0249-z.

Shah, S., O'Connell, P., Hosking, J., 1996. Modeling the effects of spatial variability in rainfall on catchment response. 2. experiments with distributed and lumped models. J. Hydrology 175, 89–111.

Shimizu, K., 1993. A bivariate mixed lognormal distribution with an analysis of rainfall data. Journal of Applied Meteorology 32, 161171.

Sklar, A., 1959. Fonctions de répartition à n dimensions et leurs marges. Publications de l'Institut de Statistique de L'Université de Paris 8, 229–231.

Sklar, A., 1996. Random variables, distribution functions, and copulas a personal look backward and forward, distributions with fixed marginals and related topics. ed. by L. Rüschendorff, B. Schweizer, and M.D. Taylor, Hayward, CA, Institute of Mathematical Statis, 1–14.

Smith, J., DeVeaux, R., 1992. The temporal and spatial variability of rainfall power. Environmetrics 3, 2953.

Smith, J., Krajewski, W., 1993. A modeling study of rainfall ratereflectivity relationships. Water Resources Research 29, 25052514.

Smith, J., Shed, R., Waltson, M., 1989. Parameter estimation for the nexrad hydrology sequence. Preprints, 24nd Conference on Radar Meteorology, American Meteorological Society, Tallahasee, FL, 259-263.

Sorooshian, S., Gupta, V., Fulton, J., 1983. Evaluation of maximum likelihood parameter estimation techniques for conceptual rainfall-runoff models: Influence of calibration data variability and length on model credibility. Water Resources Rearch 19, 251–259.

Spearman, C., 1904. The proof and measurement of association between two things. Amer. J. Psychol. 15, 72101.

Staff, W. Q., Laboratory, W. R., 1983. Hydrology, erosion, and water-quality studies in the southern great plains research watershed, south-western oklahoma, 1961-78. U.S. Dept. Agric. Agric. Res. Serv. Rep. ARM-S-29.

Steiner, M., Smith, J., 2000. Reflectivity, rain and kinetic energy flux relationships based on raindrop spectra. J. Appl. Meteorol. 39, 1923–1940.

Steiner, M., Smith, J., 2002. Use of three-dimensional reflectivity structure for automated detection and removal on non precipitating echos in radar data. J. Atmos. Oceanic Technol. 19, 673–686.

Stout, G., Mueller, E., 1968. Survey of relationships between rainfall rate and radar reflectivity in the measurement of precipitation. Journal of Applied Meteorology 7, 465474.

Syed, K., Goodrich, D., Myers, D., Sorooshian, S., 2003. Spatial characteristics of thunderstorm rainfall fields and their relation to runoff. Journal of Hydrology 271, 121.

Tetzlaff, D., Uhlenbrook, S., 2005. Significance of spatial variability in precipitation for process-oriented modelling: results from two nested catchments using radar and ground station data. Hydrology and Earth System Sciences 9, 29–41.

Tian, Y., Peters-Lidard, C., Eylander, J., Joyce, R., Huffman, G., Adler, R., Hsu, K., Turk, F., Garcia, M., Zeng, J., 2009. Component analysis of errors in satellite-based precipitation estimates. J. Geophys. Res. 114 (D24101), doi: 10.1029/2009JD011949.

Troutman, B. M., 1983. Runoff prediction errors and bias in parameter estimation induced by spatial variability of precipitation. Water Resources Research 19 (3), 791–810.

Ulbrich, C., 1978. Relationships of equivalent reflectivity factor to the vertical fluxes of mass and kinetic energy of hail. Journal of Applied Meteorology 17, 18031808.

Venkatesan, C., Raskar, S., Tambe, S., Kulkarni, B., Keshavamurty, R., 1997. Prediction of all india summer monsoon rainfall using error-back-propagation neural networks. Meteorology and Atmospheric Physics 62, 225–240.

Venter, G., 2002. Tails of copulas. In: Proceedings of the Casualty Actuarial Society, LXXXIX 2002. p. 68113.

Villarini, G., Krajewski, W., Ciach, G., Zimmerman, D., 2009. Product-error-driven generator of probable rainfall conditioned on wsr-88d precipitation estimates. Water Resources Research 45 (1), W01404.1–W01404.11, doi: 10.1029/2008WR006946.

Villarini, G., Serinaldi, F., Krajewski, W., 2008. Modeling radar-rainfall estimation uncertainties using parametric and non-parametric approaches. Advances in Water Resources 31 (12), 1674–1686, doi: 10.1016/j.advwatres.2008.08.002.

Wilks, D., 1998. Multisite generalization of a daily stochastic precipitation generation model. J. Hydrology 210, 178–191.

Wilks, D., Wilby, R., 1999. The weather generation game: A review of stochastic weather models. Progress in Physical Geography 23 (3), 329357.

Wilson, J., Brandes, E., 1979. Radar measurement of rainfall: a summary. Bull. Am. Met. Soc. 60, 1048–1058.

Winchell, M., Gupta, H., Sorooshian, S., 1998. On the simulation of infiltration and saturation-excess runoff using radar-based rainfall estimates: effects of algorithm uncertainty and pixel aggregation. Water Resources Research 34 (10), 26552670.

Xuan, Y., Cluckie, I., Wang, Y., 2009. Uncertainty analysis of hydrological ensemble forecasts in a distributed model utilising short-range rainfall prediction. Hydrology and Earth System Sciences 13 (3), 293–303.

Yilmaz, K., Hogue, T., Hsu, K., Sorooshian, S., Gupta, H., Wagener, T., 2005. Intercomparison of rain gauge, radar and satellite-based precipitation estimates with emphasis on hydrologic forecasting. Journal of Hydrometeorology 6 (4), 497517.

Young, C., Bradley, A., Krajewski, W., 2000. Evaluating nexrad multisensor precipitation estimates for operational hydrologic forecasting. Journal of Hydrometeorology 1 (3), 241–254.

Zawadzki, I., 1984. Factors affecting the precision of radar measurements of rain. Preprints, 22d Conf. on Radar Meteorology, Zurich, Switzerland, American Meteorplogical Society, 251256.

Zhang, L., S., R., Singh, V., 2008. Bivariate rainfall frequency distributions using archimedean copulas. Journal of Hydrology 32 (1-2), 93–109.

Zhu, C., Park, C., Lee, W., 2008. Statistical downscaling for multi-model ensemble prediction of summer monsoon rainfall in the asia-pacific region using geopotential height field. Advances in Atmospheric Sciences 25 (5), 867–884.

Made in the USA
Monee, IL
05 November 2026

67811179R00105